# Praise for CONVERGENCE

*Convergence: The Interconnection of Ext~* ~per-
*iences* is an exceptional book tha⁺ ss!
Authors Barbara Mango and ⁻ 'r
writings about consciousne. ...s.
This visionary book makes a ~ non-local
consciousness may be the fₒ ₒr all anomalous
phenomena. Supporting this grₒ ...d-breaking conclusion is
extensive evidence from leading researchers and the amazing
personal experiences of the authors. This is the book for you
whether you are exploring your personal exceptional
experience or interested in the latest evidence regarding the
interconnection of non-local consciousness phenomena. This
outstanding book is expertly written, beautifully easy to read,
and recommended enthusiastically.

- Jeffrey Long, M.D., author
of the New York Times bestselling *Evidence of the Afterlife:
The Science of Near-Death Experiences*

What I love about *Convergence: The Interconnection of
Extraordinary Experiences* is the union of the subjective
experience and the objective data. As someone intimately
familiar with near-death experiences, mediums, and even
ufology, I see the great need to shine more honest, insightful
light into these areas. Skeptical debunkers have an agenda that
severely limits their ability to honestly accept certain valid and
ever-growing data. This book will hopefully open the minds of
traditional scientists and skeptics alike.

- Ned Matinnia, IANDS Board Member,
Spiritually Transformative Experiencer, and NDE Advocate.

Barbara Mango and Lynn Miller contend 'anomalous' phenomenon, like love, is only fully comprehended via experience. It is a perspective which echoes my personal understanding and research into this phenomenon. The authors courageously and honestly share their 'multi-dimensional 'experience, and how it had a deep and profound impact on their lives, understanding of themselves and the nature of reality. What does it mean to be a conscious sentient being experiencing expanded awareness and alternate realities? This book offers an invaluable insight for the curious, the explorer, but most particularly a validating resource, for those who walk the road less travelled. Highly recommended.

- Mary Rodwell, Co-founder and Principal
of Australian Close Encounter Resource Network (ACERN),
author of *Awakening* and *The New Human*, Co-Founder
and Board Member of the Edgar Mitchell FREE Foundation

This book is chock full of uplifting messages and inspiring stories for anyone who's had spiritual experiences. Messages like "Shift happens!" "You're not crazy!" "The plant won't grow until the seed is cracked," or "What appears to be a curse is often really a blessing.

- Robert Peterson, Author, *Out-of-Body Experiences:
How to Have Them and What to Expect*

# CONVERGENCE

## The Interconnection of
## Extraordinary Experiences

**Barbara Mango, PhD**
**Lynn Miller, MS**

**Foreword by Erica McKenzie, RN**
*Penny Wilson, Contributor*

*atmosphere press*

# Contents

*Preface* **by Barbara Mango, PhD**                    3
*Foreword* **by Erica McKenzie, RN**                   7
*Acknowledgments*                                      10

**Chapter 1: INTRODUCTION**                            13
**Chapter 2: EMBRACING OUR UNIQUENESS: THE**
**            ANOMALOUS-PRONE PERSONALITY**            21
**Chapter 3: AM I INSIDE OR OUT OF MY MIND?**          44
**Chapter 4: DIVIDED MINDS: SKEPTIC VS**
**            EXPERIENCER**                            60
**Chapter 5 : CONVERGENCE**                            88
**Chapter 6 : IT'S A MIRACLE! SPONTANEOUS**
**            HEALING**                                160
**Chapter 7: THE NEW ME: TRANSFORMATIVE**
**            AFTEREFFECTS**                           180
**Chapter 8: WEIGHING IN: OUR STORIES**                207
**Chapter 9: CHANGE IS IN THE AIR**                    239
**Chapter 10: FINAL THOUGHTS**                         259

*Resources*                                            265
*Endnotes*                                             270
*Bibliography*                                         302

# Preface

Barbara Mango, PhD

*When I say work, I only mean writing.*
*Everything else is just odd jobs.*
Margaret Laurence

*Every secret of a writer's soul, every experience of his life,*
*every quality of his mind, is written large in his works.*
Virginia Woolf

I believe the account of *how* a book came to be written is often as interesting as the narrative itself. Writing a book is alternately exhilarating, incredibly stimulating, frustrating, and at times, just plain overwhelming. Of course, the proverbial "writer's block" rears its ugly head at the most unwelcome of times. Many days my fingers flew effortlessly over the keyboard, the inspirational juices flowing, while others, I stared aimlessly at a blank word document. Ah, the life of a writer.

Completing a book by a single author is challenging. However, it is much more complex to write a book with two co-authors and a contributor, living hundreds of miles apart. Yet, Penny, Lynn, and I combined our strengths and compatible personalities to create a finished work. The three of us are lifelong experiencers who share the same delightfully wicked sense of humor, own a multitude of misbehaved felines, and have a passion to investigate consciousness and its connection to extraordinary experiences.

I met Penny at the annual Past-Life Research Institute Conference, in Los Angeles, CA. Penny and I later collaborated on *The Transformative Power of Near-Death Experiences*, written by world-renowned near-death researcher, Dr. Penny Sartori. Lynn and I served on an academic research committee investigating the connectivity between anomalous experiences and consciousness. The three of us formed a working relationship via countless Skype and phone conversations. Most importantly, however, we became close friends.

*Convergence, the Interconnection of Extraordinary Experiences* originated from an article written by Lynn and me. The two of us have long wished to write a collaborative book— one which examines extraordinary phenomena from an *experiencer's* perspective, yet, is also backed by science. Science is an essential component in examining consciousness and its interconnection with paranormal phenomena. Its methodology includes testing hypotheses, gathering evidence, experimentation and or observation, critical analysis, verification, empirical testing, and peer review.

Thus, science requires hard data to "prove" that which experiencers claim to "just know." After all, they are the ones undergoing *experiential* encounters. These are profoundly real, rather than conceptual occurrences. Experiencers routinely refer to them as "realer than real." And how can they not be? During such experiences, individuals report heightened senses, greater than 360% vision, telepathic communication, and inexplicable medical and phobic healings.

Lynn, Penny, and I are lifelong experiencers. Due to a condition termed idiopathic anaphylaxis, Penny has undergone four near-death experiences. Barbara is a lifelong experiencer of "all things paranormal," including recall of numerous past lives. Since childhood, Lynn has experienced hundreds of spontaneous and controlled out-of-body experiences, in addition to multiple Unidentified Aerial Phenomena (UAP) related contact.

We have found these phenomena alternately transcendent, confusing, and wondrous—each a journey beyond space and time. They have transformed our lives in profound ways, impacting relationships, reconfiguring our world view, and giving us a peek into the mysterious workings of non-local consciousness. We believe we have "lived" what science is attempting to "prove." After all, we are human beings having experiences which lay beyond our three-dimensional, everyday reality.

*Convergence, the Interconnection of Extraordinary Experiences* examines the concept that non-local consciousness may be the foundation of all anomalous phenomena. Although there are numerous otherworldly occurrences, our focus is on four specific areas: Unidentified Aerial Phenomena related contact (UAP), Near-Death Experiences (NDEs), Out-of-Body Experiences (OBEs), and Past-Life Recall/Past-Life Regression, (PLR, PLRT). Each area is examined from both an experiential and scientific viewpoint.

The authors propose all such experiences are interconnected, rather than independent modalities. Our hope is that you, the reader, find this book "different" and unique. Lynn, Penny and I are multiple and lifetime experiencers. Collectively, we have taken nearly 1,000 otherworldly journeys. We have personally experienced every phenomenon discussed in this book. Our backgrounds/training include biology, psychology, nursing, metaphysics, and education. We are seekers, with insatiably curious minds.

During the writing process of this book, the authors had several objectives in mind. We chose to examine these specific four phenomena for four reasons. First, and foremost, was our hope that all who read this book do so with open hearts and minds. To be open to the concept that *experience* holds as much weight as the science behind it. To understand that those of us who came into this world "wired differently" are not weird or crazy, but non-pathological, highly sensitive, and uniquely gifted individuals. Secondly, all topic areas have been meticulously researched by world-renowned physicians, psychiatrists, psychologists, neurologists, physicists, philosophers, ufologists, academics, and quantum theorists. Thirdly, the authors have personally experienced *every* experiential modality discussed in this book. Finally, we explore the inherent personality traits which makes one *prone* to having extraordinary experiences. Why are some individuals seemingly predisposed to encountering inexplicable phenomena, while others are not?

Our experiences have demonstrated (to us) that extraordinary phenomena are interconnected via non-local consciousness. Our otherworldly journeys have revealed the numerous commonalities which exist among these occurrences. Our lives, as those of nearly every experiencer, have been transformed forever.

The author's personal experiences are interspersed (and injected with wee bits of humor) throughout this book. Renowned experts in the fields of UAP, OBEs, PLRT (Past-life Regression Therapy), NDEs, neuroscience, and paranormal phenomena contribute their personal experiences and perspectives. These include Dr. Bob Davis, internationally recognized neuroscientist; Dr. Heather Friedman Rivera, author and co-founder of the Past Life Research Institute; Bob Peterson, OBE and parapsychology researcher; Erica McKenzie, RN, author, and near-death experiencer; Marcie Klevens, MA, LMHCA, and UFO-contact/abductee experiencer; and Brent Raynes, paranormal/UFO investigator, and editor of *Alternate Perceptions* magazine.

We have provided links for several personality inventories that are simple to take, fun, and best of all, free. We have additionally furnished a list of available support groups, forums, and hypnotherapists. For too long, experiencers have remained in the closet, afraid to share their stories for fear of ridicule, professional suicide, or accusation of mental disorders. Let's begin a new dialogue—one in which we "reframe" the way we view ourselves, not as odd or delusional, but uniquely gifted. There is a global community of like-minded individuals to connect with in a safe and supportive environment. They are merely a few keystrokes away.

6

# Foreword

By Erica McKenzie, RN

The majority of medical school curriculum continues to teach the materialistic model.

Materialists believe that the entire universe is unconscious; composed of matter and mathematical laws. This model contends that consciousness is confined to the physical brain. Therefore, the idea of paranormal phenomena cannot be explained by scientific means. Yet, I've lived my entire life co-existing with the paranormal, in a world where boundaries do not exist between this life and the next. In fact, this way of life is my "normal."

My research suggests that altered states of consciousness enable us to access supernatural dimensions of reality. These dimensions include but are not limited to: Un-identified Aerial Phenomena (UAP), Near-Death Experiences (NDEs), Out-of-Body Experiences (OBEs), and Past Life Recall (PLR). Since 2002, I've conducted countless interviews with people from all walks of life who have claimed experiences, including but not limited to a feeling of oneness with the universe and perceiving other dimensions of reality beyond Earth.

Generally, our society labels individuals who have extraordinary experiences as psychologically unstable. Yet, the numbers of mentally sound individuals reporting anomalous phenomena are steadily increasing. After such experiences, the vast majority of people report positive transformation. Most extraordinary of all, these individuals have found their sense of purpose and value.

Historically, numerous parallels exist in stories documented by saints, shamans, mystics, and yogis. Spiritual writings across time, tradition, civilization, and culture corroborate the healing and transformative power of these extraordinary experiences.

*Convergence: The Interconnection of Extraordinary Experiences* explores phenomena which proposes to change the very foundation of societies' beliefs. Hopefully, this may change how we view our very reality.

A dear friend of mine, physicist Dr. Claude Swanson, said, "I've pursued investigations into 'unconventional physics.' My principal interest has been unified field theory, the so-called 'Theory of Everything' which could explain the universe at the deepest possible level. This has led me to investigate many aspects of the paranormal, which appear to be completely real phenomena, which violate our present science. Paranormal phenomena, which have now been proven in the laboratory in many cases, offer a window into the deeper universe, the mysteries of consciousness, and unlock new forces and principles which conventional science has only begun to glimpse."

Claude's research indicates that out-of-body and near-death experiences demonstrate that other dimensions exist. Paranormal research suggests that the human soul exists outside the body in the form of energy and is the center of human consciousness. If it is true that we are far more than our physical bodies, as the late scientist and pioneer Robert Monroe once stated, then it is imperative to keep an open heart and mind when reading this book.

I believe the information in this book is crucially important in understanding non-locality and the interconnection of non-ordinary phenomena. *Convergence: The Interconnection of Extraordinary Experiences* explores reality beyond the brain, not limited to the after-life, but as the path that leads to the connection of all extraordinary phenomena. What sets this book apart from others is:

- The authors are lifelong experiencers and academic professionals, with credentials in meta-physics, science, education, and psychology.
- The authors have personally experienced all

modalities explored in this book.

- The authors combine their own personal experiences, case studies, and scientific research.
- The authors emphasize the *experiential* component of extraordinary experiences, which they consider as significant as scientific data.
- The authors honor the experiencers, their stories, and their truths. As experiencers themselves, they are compassionate and empathetic to the reality and integration of extraordinary phenomena.
- The authors recognize experiencers as unique and gifted individuals.
- The authors have the unique ability to present profound scientific concepts in an easily understandable, relatable manner.

*Convergence: The Interconnection of Extraordinary Experiences* is a truly inspirational book. It is only when we open our minds and hearts to embrace experiences which lay beyond our five senses that we may begin to explore the very nature of consciousness.

# Acknowledgments

To Nick Courtright, Executive Editor, Atmosphere Press. An enormous thank you for your belief in us. Your humor, willingness to take on our "unusual" topic, and author-centric approach has been amazing. To our wonderful editor Alex Kale, I have valued your keen insight, suggestions, and kindness more than you know. It has been a joy to work with you. Penny Wilson, thanks enormously for your wonderful, and very personal contributions to this project. I would like to thank my husband, Greg, for always believing in me and letting me fly free. Bob Davis, Brent Raynes, Heather Friedman Rivera, Erica McKenzie, Marcie Klevens, and Bob Peterson, much appreciation for sharing your insightful journeys. To Jeffrey and Jody Long, who not only published my first post PhD articles on their NDERF website, but also connected me with individuals who would forever change my life. A big shout-out to my friend Erica McKenzie, who was instrumental in helping me to embrace my uniqueness—it means the world and more. To Lynn, your friendship and shared love of humor, art, food, and felines is valued more than you know. To everyone who constantly cheered me on: Sandy, Amy, Jess, Laura, Brent, Alisa, Scott, Aiden, and Sophie. Lynn and Penny, what a wild adventure this has been. To all aspiring authors, never let fear get in the way of your dreams, and never give up.

Barbara Mango, PhD

First and foremost, thanks to Atmosphere Press; your support has been invaluable. A big thank you to my beautiful daughter, who put forth so much effort into the graphics and for her support through all of this. I would like to thank the wonderful contributors for taking the time to open your hearts to us. There are so many things that happened in my life that have brought me to the place I am now. I have faced numerous hardships and challenges. Yet, without these, I would have been unable to realize my inner strength and purpose, and I would not have had the strength to climb out of the trenches and stand tall back on my feet. To William Buhlman, author, researcher, and educator of Out-of-Body Experiences. You, sir, have rocked my world and completely changed my paradigm. To Claudia Lambright, moderator of William Buhlman's forum. You have taught me to both understand my experiences and grow exponentially. To Teri Zeman, the first person I shared my experiences with in-person, who did not judge me or think I was insane. To the individual who put me on my soul's path, Reinerio Hernandez, co-founder of FREE (The Doctor Edgar Mitchell Foundation for Research into Extraterrestrial and Extraordinary Encounters). You changed my life by opening up a whole new world for me, and by introducing me to people who have become my tribe, my friends. You introduced me to someone who not only became the co-author of this book, but my friend, my mentor, my advisor, counselor, and so many more things. Finally, to the co-author of this book, Dr. Barbara Mango, I never thought a friendship could bloom so huge. You have been there for me through good times and especially the bad, with such complete understanding and unconditional love. Thank you, my dear friend.

Lynn Miller, MS

# Chapter 1
## Introduction

*It is up to each and every spirit to decide "how far down the rabbit hole" we want to go to discover our infinite possibilities and capabilities.*[1]
Michael J. Robey, Psychic Medium

Is the physical world all there is, or could there be something more? Perhaps the physical realm is simply a minute fragment of a much larger existence. The vast multitudes of individuals who have reported anomalous experiences beg us to explore the possibility that we live in a multi-dimensional, rather than 3D reality. Extraordinary experiences such as NDEs (Near-death experiences), OBEs (Out-of-body experiences), UAP-related contact (Unidentified Aerial Phenomena), and PLM (Past-life memories) have been recorded since time immemorial.

Approximately ten thousand years ago, primeval civilization first recorded otherworldly phenomena and out-of-body journeys. Prehistoric man drew images of spacecraft and non-human beings on cave walls. Plato's legendary manuscript, *The Republic*, (circa 380 BCE) examined the first account of a NDE in Western Civilization. This Socratic dialogue illustrated Plato's belief that the soul, or incorporeal body, separates from the material body upon death. Approximately 2,600 years ago the first reference to reincarnation appeared in the Upanishads (the sacred scriptures of Hinduism).[2]

*Image 1-1*
*Illustration of Egyptian Hieroglyphic 2,000 - 3,000 BCE,*
*depicting Ancient Astronauts*

Throughout recorded history, accounts of anomalous experiences span civilization, culture, and tradition. These earliest accounts describe elements that are remarkably parallel to case studies examined in contemporary research. Thus far, mainstream science has consistently dismissed the validity of such phenomena, writing it off as hallucinations, confabulations, fantasies, oxygen deprivation, or chemical reactions of a dying brain. Yet, it is only through careful consideration and study of these claims that we can determine if something much larger is going on. By examining each case, its details, and common themes, we open a broader dialogue into consciousness and what it entails. We also begin to probe whether consciousness is local—that is, a manifestation of brain function—or a non-local phenomenon, *accessible, but not confined,* to the brain.

*What is mind? No matter. What is matter? Never mind.*[3]
Scholar Thomas Hewitt Key

Cutting-edge research into quantum mechanics (a theory describing the world of atoms and subatomic particles and their interaction between energy and matter) has made it increasingly difficult for mainstream science to outright reject the claims of experiencers. A growing body of evidence suggests we exist in a multidimensional universe in which all anomalous phenomena are inter-connected, via non-locality. Non-locality contends that consciousness is independent of space and time, suggesting the brain functions as an intermediary between spirit and body. As lifetime experiencers, Lynn and I contend that NDEs, OBEs, UAP-related contact, and PLMs are indeed interconnected. Furthermore, we maintain non-locality is the *very vehicle* through which these phenomena are experienced.

*Image 1-2*

*Illustrative Representation of Non-local Consciousness*

The field of conscientiology (originally proposed in 1986 by professor and researcher Waldo Vieira) theorizes, as do we, that consciousness is multidimensional in nature. The fundamental principles of conscientiology assert:

Consciousness is multidimensional, that is, it manifests itself in several dimensions.

Consciousness is set in multiple vehicles (bodies) used by the consciousness to manifest itself in different dimensions: the soma (physical body) for the physical dimension, the energosoma (chakra, energy body, etheric double) for the ethereal dimension, the psychosoma (astral body, spirit body) for the non-physical dimension, and the mentalsoma (mental body) for the mental dimension.

Consciousness is multi-existential, meaning it has existed before physical birth and will continue to exist after death. After the extinguishment of the physical body, consciousness will manifest in the extra-physical dimensions, employing its other bodies. In fact, it alternates in a serial fashion between periods in the physical dimension (lives) and periods in the extra-physical dimension.

Consciousness is an evolutionary process. Manifestation from the most basic to the most complex forms of life continues to change and adapt. Conscientiology not only identifies and studies the different levels along the scale of evolution, but also how consciousness evolves more efficiently through its serial existences. Evolutionary advancement is a primary driving force of consciousness.[4]

From inexplicable journeys beyond time and space, experiencers appear to implicitly understand that consciousness is independent of the brain. World-renowned cardiologist and near-death researcher Pim van Lommel illustrates the concept of non-locality:

Our brain may be compared both to a television set, receiving information from electromagnetic fields and decoding this into sound and vision, and to a television camera, converting or encoding sound and vision into electromagnetic waves. Our consciousness transmits information to the brain and via the brain receives information from the body and senses. The function of the brain can be compared to a receiver; our brain has a facilitating rather than a producing role; it enables the experience of consciousness.[5]

The late Paul Pietsch, PhD and Professor Emeritus at Indiana University, explored non-locality in a slightly more graphic manner:

Dr. Pietsch experimented with salamanders to locate where memories are stored in the brain. He removed their brains, ground them up, even shuffling their brains around, and then placed them back in their heads. The astonishing result was their memories were unaffected although their brains were demolished. Pietsch concluded that memory was not a local phenomenon, but instead, was linked to something outside their bodies. His findings were published in his book, *Shufflebrain: The Quest for the Holo-graphic Mind*.[6]

In 2005, the journal *Science* published an issue listing the top 125 questions scientists had yet to answer. The most significant inquiry was, "What is the universe made of," immediately followed by, "What is the biological basis of consciousness?" This question may be reframed by asking, "Does consciousness have a biological basis at all?"[7]

Researchers in the fields of psychology, neuroscience,

philosophy, physics, and parapsychology seek answers to these questions. Yet, is there, or will there ever be a theoretical model which can fully "prove" this deeply profound concept? Countless books, academic research, and statistical analyses have examined such phenomena, but they have yet to come up with a definitive theory. Perhaps the only way consciousness can be comprehended is experientially. Biochemically, "love" may be rationalized as a mere cocktail of brain chemicals: dopamine, serotonin, oxytocin, and endorphins. Yet, love is infinitely more complex than mere neuroscientific analysis of the brain. Falling deeply in love or holding a newborn infant in one's arms cannot be rationalized by science. Love, an intense, overwhelmingly powerful emotion, must be *experienced*, rather than *explained* via physiological/ neurological analysis.

Thus, the authors contend that anomalous phenomenon, like love, is only fully comprehended *via* experience. Experiencers journey to a realm in which non-local consciousness is omnipresent or to the source of the fundamental substance of our physical existence, while being completely independent of it. This wild ride changes us forever—how could it not? Our experiences are embedded into our very being and become our stories, our lives, and our truth. We are not implying that science is not fundamentally significant or invaluable to society. After all, without science, our modern world would not be modern, and our understanding of it would remain archaic. It has been stated that "Accepted scientific ideas are reliable because they have been subjected to rigorous testing, but as new evidence is acquired and new perspectives emerge, these ideas can be revised." Thus, we must consider *why* traditional science remains unwilling to consider "new perspectives."[8]

> The modern scientific worldview is predicated on assumptions that are strongly associated with classical physics. Materialism,

the idea that matter is the only reality, is one of these assumptions. However, the nearly absolute dominance of materialism in the academic world has seriously constricted the sciences and hampered the development of the scientific study of mind and spirituality. Faith in this ideology, as an exclusive explanatory framework for reality, has compelled scientists to neglect the subjective dimension of human experience. This limited reasoning has contributed to a severely distorted and impoverished understanding of ourselves and our place in nature.[9]

How interesting that in a world full of beautiful mysteries, where so many things defy explanation; traditional scientists, in their arrogance, think they have it all figured out.

Penny Wilson

Hundreds of books, academic papers, and scientific findings on anomalous phenomena are published each year. Yet, relatively few of these publications are written by scientists, psychologists, neuroscientists, or academics, who additionally, are experiencers themselves. Alternatively, numerous non-scientific, non-academic individuals have written powerful stories of their own magnificent experiences (NDEs, OBEs, UAP-related contact, and reincarnation/past-life memories).

Our experiences have demonstrated to us that extra-ordinary phenomena interconnected via consciousness are non-local in nature. Our otherworldly journeys have revealed the numerous commonalities which exist among phenomena. Our lives, as those of nearly every experiencer, have been transformed forever.

In nearly all cases, individuals report the most significant impact of their experience is the profound and life-altering after-effects. Although it may take years to assimilate such intense occurrences, they are, ultimately, positive in nature. Life is viewed from a new, wider, and liberating perspective. Experiencers find themselves forming friendships, partnerships, and professional collaborations with like-minded individuals. Collectively, albeit gradually, a new mindset is setting the foundation for global transformation; a new worldview in which humanity works collectively and for a higher purpose.

Our intention is to stimulate conversation, deliberation, and lively debate about this paradigm shift. As writer, researcher, and cognitive scientist Daniel Dennett boldly states, "When scientists muster the courage to face this evidence unflinchingly, the greatest superstition of our age—the notion that the brain generates consciousness or is identical with it—will topple. In its place will arise a non-local picture of the mind."[10]

# Chapter 2
## Embracing Our Uniqueness

*The Anomalous-Prone Personality: The ASP, HSP, and FPP*

**Anomalous:** Inconsistent with or deviating from what is usual, normal, or expected: Irregular, Unusual.[1]

*Image 2-1*
*Illustration of Albert Einstein – Physicist and "Father" of the Theory of Relativity*

How many of us have lived our lives feeling "different," like we don't fit in? It's tough going through life having experiences that are considered weird or unusual. Many of us question our "normality," or perhaps, even sanity. Yet, research indicates that anomalous-prone personality types are not crazy or pathologically disturbed individuals. Rather, those possessing such characteristics are predisposed to have extraordinary experiences. These individuals frequently report experiencing unique phenomena and feeling "different" from others from an incredibly young age.

From the time I was verbal, I knew I was different. What normal two-year-old comprehends that other dimensions exist beyond space and time? I received this information from a loving, non-human entity who frequently visited me. I don't remember our conversations in detail; however, I know they concerned space-time, the cosmos, and the meaning of existence. I spoke to beings only I could see and remembered past lives. I knew, unequivocally, that Earth was not my "real" home and merely a 3D construct. My mother once said to me, "If I didn't remember giving birth to you, I would never think you were mine. You are so different from the rest of us." I will never forget that comment. Throughout my childhood, I felt like an unloved and misunderstood "freak."

Barbara

From an incredibly early age, I had memories and visions of tunneling under the ground towards a huge structure. In this underground chamber were objects of great power and importance. Later, as I became more aware of the world, I realized that this large structure was a Mayan Pyramid. I never "fit in" with others. I spent most of my time alone, exploring nature and my surroundings, looking for snakes and lizards. It wasn't until about sixth grade that I started to make friends and became more social. It's weird, because I

22

never really thought about this until I was asked to recall my early childhood. My real friends were animals and my pets.

Lynn

The world has always seemed a malleable place for me. As a child, I thought everyone could drift off into that "in-between" place. It is so easy to pop out of my body between the waking and sleeping states. I flew all over our neighborhood and even saw items on rooftops that were later proven to be there. I sensed when someone was going to die. I received a phone call from my grandmother, after her death! My mother was standing next to me when I answered the call. It was a strange, static-filled conversation. My grandmother asked to speak to my mom. I handed mom the phone, as though there was nothing strange about it at all. My poor mother nearly fainted at hearing my deceased grandmother's voice. These were normal happenings in my life, and not until my teen years, did I realize that everyone else wasn't having similar sorts of occurrences.

Penny

Personality types in both near-death and contact experiencers have been extensively studied by Dr. Kenneth Ring. A summary of his research, entitled "The Omega Project: An Empirical Study of the NDE-Prone Personality" was published in the *Journal of Near-Death Studies* in 1990. Ring purports that elements in childhood predispose certain individuals *to either have or recall having a NDE*. His later research identified an "encounter-prone" [*anomalous*] personality type—a distinctive, spiritually sensitive and ecologically-oriented individual who collectively, may represent the next stage in human evolution.[2]

Encounter-prone personalities consistently report traumatic events (i.e., accidents, injuries, illnesses, and/or psychological shock) and/or abusive childhoods (including physical, psychological, or sexual abuse), a negative home environment, susceptibility to experiencing non-pathological alternative realities, and mild dissociation.[3] It is important to emphasize that the dissociation associated with the anomalous-prone personality is *non-pathological* in nature. In fact, mild dissociation is experienced by most stable, emotionally healthy adults. For example, how many of us have driven on the highway, not recalling the exits we passed while "zoning out?"

Nevertheless, traditional medicine continues to "diagnose" ASPs with pathological, rather than mild dissociation. There are three types of pathological dissociative disorders defined in the DSM-5:

> ***Dissociative amnesia-*** Difficulty remembering important information about one's self. Dissociative amnesia may surround a particular event, such as combat or abuse, or more rarely, information about identity and life history.
>
> ***Depersonalization disorder-*** Ongoing feelings of detachment from actions, feelings, thoughts, and sensations, as if one is watching a movie (depersonalization). Sometimes other people and things may feel like people and things in the world around them are unreal (derealization). A person may experience depersonalization, derealization or both.
>
> ***Dissociative identity disorder-*** Formerly

24

known as multiple personality disorder, this disorder is characterized by alternating between multiple identities. A person may feel like one or more voices are trying to take control in their head. Often these identities may have unique names, characteristics, mannerisms, and voices. People with DID will experience gaps in memory of everyday events, personal information, and trauma.[4]

Dr. Frank W. Putnam, a professor of psychiatry with a specialization in pathological dissociation, emphasizes that non-pathological dissociation is an "intelligent and creative survival strategy" which serves several major functions, including efficiency and economy of effort, resolution of irreconcilable conflicts, cathartic discharge of certain feelings, and the submersion of the individual ego for the group identity."[5] Therefore, dissociation may serve as a "bridge," linking cognition with anomalous phenomena. It has been proposed that mild dissociation associated with the ASP, HSP, or FPP is considered non-pathological.

It is literally impossible for me to drive from point A to point B without doing what I refer to as "zoning out." I try to focus, I really do, but then off I go again. The same thing happens during conversations. Unless I am engaged in a truly fascinating discussion, my mind just wanders anywhere and everywhere. I feel badly that I am not giving the other individual my full attention. I imagine that my eyes just glaze over or get a distant or blank look. Yet, surprisingly, no one ever seems to notice I'm "spacing out." Am I mildly dissociating? Absolutely. This most likely explains why I can "flip into" other dimensions so easily.

Barbara

This is one of my most prominent traits that I would not consider to be mild. Not only has it been an adaptive response to stress, but also to boredom. Throughout my life, I have been so intensely engrossed in my fantasy world that my lips and sometimes arms would move as I would play the scenario in my head. It is for this reason that I have often longed to write fantasy and sci-fi novels.

I dissociate in my car and my daughter often comments, "Are you talking to yourself again?"

Lynn

Dissociation—I think we all do it to some degree. Who hasn't gotten caught up in a movie and found himself on the edge of his seat? It's a bit more than that for me though. From a very young age, I could sort of "float out" of my physical body. During waking hours, I didn't go far, generally just losing awareness of my physical self. In the moments before sleep, I drifted out of my physical form and went on grand journeys through both my town and other countries. It wasn't like dreaming, as it was so lucid, and I remember those travels to this day. As an adult, I have to remind myself to "stay in," or, as a friend of mine calls it, "stay in all the way to my toes." Otherwise, I tend to lose awareness of my surroundings and drift off to other places. I've driven as far as 60 miles with very little recall of the journey. This goes beyond daydreaming and I work to keep it in check. I've never had an accident, so some part of my mind is definitely present, doing the driving and minding traffic, but that other part of me, the part that likes to fly away, seems to have a mind of its own.

Penny

Studies have identified three anomalous-prone personality types: the ASP, HSP, and FPP. Although there are numerous similarities and overlap among the three, certain distinctions remain.

### The ASP: Anomalously Sensitive Person

An Anomalously Sensitive Person exhibits uncommonly high levels of sensitivity, not only in the emotional realm, but also in the physiological, cognitive, altered states of consciousness, and transpersonal ("meta-physical") realms as well. With highly attuned emotional sensitivities, it is not surprising that ASPs frequently report coexisting issues, including physiological, autoimmune, biochemical, sensory, psychological, electromagnetic, and environmental sensitivities.

Psychiatrist James Lake defines anomalous personality types in this way: "Characterized by a propensity to have uncommon or unusual experiences including paranormal experiences, mystical experiences, and transpersonal experiences." Lake additionally suggests that an anomaly-prone personality "May be associated with increased neural and cognitive connectivity, as well as relatively greater creativity and increased vividness of mental imagery."[6]

According to David Ritchey, researcher and author of *The H.I.S.S. of the A.S.P.: Understanding the Anomalously Sensitive Person*, more than five million Americans are

Anomalously Sensitive.

Ritchey contends that ASPs score higher on the HISS (Holistic Inventory of Stimulus Sensitivities) Questionnaire. "HISS survey respondents score high on scales of intelligence, imaginativeness, assertiveness, forthrightness, venturesomeness, self-assuredness, and self-sufficiency. Additionally, they demonstrate an inclination to be creative, curious, insightful, incorruptible, iconoclastic, non-conforming, and anti-authoritarian." [7]

A higher than normal percentage of ASPs are left-hand dominant. Approximately ten percent of the world's population is left-handed. Ritchey determined this prevalence is significantly higher in both HSPs/ASPs. According to his research, 20% to 30% of all subjects reported left-hand dominance, compared to 10% of the general population.[8] Left-handedness has long been associated with creativity, imagination, emotional expression, spatial awareness, and possessing the ability to "see the big picture." Many individuals associate left-handedness with left-brain dominance, when in fact, the opposite is true. In actuality, the brain is "cross-wired." The right hemisphere controls the left-handed side of the body and vice versa.[9]

*Only left-handers are in their right minds.*
*-Anonymous*

### ASP and MBTI
A significant portion of Ritchey's book investigates the correlation between anomalous personality and the Myers-Briggs Type Indicator (MBTI). Created by Isabel Briggs Myers and her mother, Katherine Cook Briggs, the MBTI is a self-guided questionnaire designed to measure one's personal and decision-making preferences. It is considered one of the most popular and widely used personality-type tests. The MBTI

recognizes sixteen distinctive personality preference types, or orientations, with four sets of contrasting variables. These include:

1. An orientation towards extraversion or introversion.
2. A perceiving strategy based on either sensation or intuition.
3. A judging strategy based on either thinking or feeling.
4. An exhibited preference for either the perceiving function or the judging function.

| ISTJ | ISFJ | INFJ | INTJ |
|---|---|---|---|
| Responsible Executors | Dedicated Stewards | Insightful Motivators | Visionary Strategists |
| ISTP | ISFP | INFP | INTP |
| Nimble Pragmatics | Practical Custodians | Inspired Crusaders | Expansive Analyzers |
| ESTP | ESFP | ENFP | ENTP |
| Dynamic Mavericks | Enthusiastic Improvisors | Impassioned Catalysts | Innovative Explorers |
| ESTJ | ESFJ | ENFJ | ENTJ |
| Efficient Drivers | Committed Builders | Engaging Mobilizers | Strategic Directors |

*Image 2-2*
*Myers-Briggs Type Indicator (MBTI). Assessment designed to pinpoint psychological preferences in how individuals perceive the world and make decisions.*

Ritchey determined that two distinct personality traits are associated with ASPs: intuition and introversion. Ritchey terms these two types the "IN-s," or introverts, with an intuitive strategy for perceiving and a feeling approach for judging. Collectively, IN-s comprise 4% of the populace. Among the IN-s, INFPs are the least common personality type, comprising a mere 1% of all people.[10] These individuals are considered the most spiritual, metaphysical and psychic of all the MBTI personality types. They value service, charity, and exploring human potential. These individuals are inclined to have a strong interest in parapsychology, occultism, and esoteric subject matter.

However, because of their unique traits, INFPs frequently often feel ostracized, misunderstood, and different (Barbara and Lynn included). Conventional individuals may feel threatened by these atypical individuals, who are intelligent, independent, non-conforming, anti-authoritarian, and unconventional. Author and Anglican Priest William Ralph Inge aptly stated:

"Public opinion [is] a vulgar, impertinent, anonymous tyrant who deliberately makes life unpleasant for anyone who is not content to be the average man." [11]

IN-s have vivid imaginations and tend to be highly artistic, musically gifted, and/or skilled writers. IN-s include noteworthy individuals such as Mahatma Gandhi, Eleanor Roosevelt, Mother Teresa, Albert Einstein, and Leonardo da Vinci.

As evidenced above, the connection between ASPs and the MBTI has been extensively examined by Ritchey. However, future research may discover a strong correlation between ASPs and the Mind Styles Model, developed by Dr. Anthony Gregorc. The Mind Styles Model identifies four preferred learning, or thinking, styles used to acquire/process information:

1. Concrete Sequential (CS)
2. Abstract Random (AR)
3. Abstract Sequential (AS)
4. Concrete Random (CR)

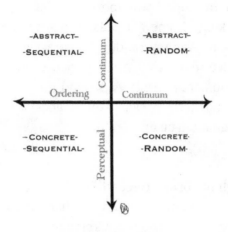

*Image 2-3*
*Gregorc Mind Styles Four Quadrant Model. Assessment determines the four preferred learning styles based on perceptual and processing modes.*

Years ago, I was accepted into an accelerated state art teaching certification program. All students were required to take the Gregorc Mind Styles Assessment in order to better understand the various learning/ thinking styles of our future students. Afterward, we were informed of our own results. (The baseline of the test results is 30 out of a total possible score of 60 points for each of the four styles. A score of 30 or higher indicates a domination in a particular style.) I remember the professor (who was, let's just kindly say, lacking in social-emotional skills) laughing aloud at my results. "It figures," he said, "Barbara is an off-the-charts Abstract Random (AR), unlike the rest of you. Her score is 58." "It figures"—what did that mean? That I was flakey? That my vivid imagination was over the top? I was mortified. Yet, it got me thinking. I knew I was a highly sensitive person. Was my random "abstract-ness" also related to my ability to experience anomalous phenomena? Did ARs and ASPs share commonalities? Oops—there goes my vivid imagination again!

Barbara

Although no formal research has been conducted on the correlation between ARs and other anomalous type personalities, there are numerous similarities. AR individuals have finely tuned perception, enabling them to "see" beyond the five senses. They have the capacity to interpret human behavior by sensing others' "vibrations." ARs are apt to question the meaning of life and/or nature of consciousness and have paranormal abilities. As children, they tend to endlessly question *everything*.

Additionally, Gregorc asserts that "ARs are able to visualize, understand, and believe that which you cannot actually see. When you are using your abstract quality, you are using your intuition, your imagination, and you are looking beyond 'what is' to the subtler implications. *It is not always what it seems.*"[12]

No one is a "pure" style. Each of us has a unique combination of natural strengths and abilities. As Elena Grigorenko and Robert Sternberg explain, "A thinking style is not a personality trait or indicator of your intelligence quotient (IQ), but an interaction of both of these elements."[13]

### The HSP: Highly Sensitive Person

The HSP has an overly sensitive nervous system, is keenly aware of subtleties in his/her surroundings, and is easily overwhelmed in highly stimulating environments.[14]

After investigating traits in ASPs, Ritchey conducted further research on HSPs. Some have capacities he calls,

"Transpersonal Experiences," which fall into three categories, and include but are not limited to:[15]

**Transpersonal Perceptions**: Deja vu, synchronicity, telepathy, precognition, psychic dreams, and clairvoyance.

**Transpersonal Influence**: Psychic healing, psycho-kinesis, and electrical psychokinesis.

**Transpersonal Manifestation of Mind**: Contact with spirit guides, out-of-body experiences, past-life recall, apparition, medium episodes, UFO sighting, near-death experiences, kundalini arousal, and alien contact.

Elaine Aron is a psychologist, psychotherapist, and highly sensitive person. This sensitivity makes her uniquely qualified to research the traits, genetic dispositions, and biology of HSPs. Aron explains that her interest in the field began with the awareness that she was *different*. She consulted a therapist after greatly overreacting to a medical procedure. The therapist suggested she might be highly sensitive.

Aron, at the time a psychologist at the University of California, Santa Cruz, became intrigued by the notion that some people might have higher levels of sensitivity than others and decided to research the subject. Aron's work led her to write the highly acclaimed book *The Highly Sensitive Person*.

Aron is convinced that there are genetic and biological bases for extreme sensitivity. As she explains:

> The brains of HSPs ... differ from those of other individuals. Studies have shown that they have more activity and blood flow in the right hemisphere of the brain, which indicates that they are internally focused, rather than outwardly oriented. The balance between the two opposing systems of the brain may account for heightened sensitivity. One system,

the "behavioral activation system," is hooked up to sections of the brain that propel people into new situations, making them curious and eager for external rewards. Another system, the "behavioral inhibition system," compares present situations to past ones before proceeding and alerts the body to be cautious in risky situations. When the behavioral inhibition system in a person's brain is the stronger of the two systems, sensitivity results.

She additionally asserts that "Sensitivity is an inherited trait that tends to be a disadvantage only at high levels of stimulation. Everything is magnified for HSPs. What is moderately arousing for most people is highly arousing for the highly sensitive. And what is highly arousing for others, is off the charts for HSPs."[16]

I have always considered myself an "anomaly." I have always been told I'm emotionally over-sensitive. I seem hyper-sensitive to everything: unnatural lighting, loud noises, hyper people, crowds, numerous odors, temperature fluctuations, you name it. I avoid malls like the plague. I'm completely overwhelmed by the commotion, crowds, and music blaring from countless stores. I feel as if I am a sponge, mopping up the emotional energy of hundreds of shoppers. My nervous system feels under attack. It's too much. The overstimulation absolutely exhausts me. I am also hyper-vigilant around others. I notice every micro facial expression, every slight shift in body language. I miss nothing. Most people consider my "aware-ness" a gift. I find it to be both a gift and exhausting curse. It may sound paradoxical that I am able to dissociate (space out), yet have extremely heightened awareness. Dissociation is an *emotional* coping method, whereas hyper-sensitivity is a *physiological* anomaly. In highly sensitive people, such as myself, these traits often co-exist.

Barbara

More and more, I want to hide from the world. It haunts me in my dreams every night. It's a double-edged sword. I am terrified of losing my job, and at the same time, I am terrified of my job. I wonder how many individuals feel this way. Maybe this is why people remain unhappy with their lives. Teachers in general often feel this way, as if the life force is sucked out of us daily. I can be strong as I walk into my classroom in the morning, and for no reason, anxiety overwhelms me. It seems to come out of nowhere, no cues or events cause this. I feel like I am somehow picking up on the feelings of teenage students. Their lives are wrapped up in the hostility of the world and each other. They are trying to find their identities. They are walking vessels of insecurity—in an internal battle, trying to find themselves, and their role in this world. I feel it, it overwhelms me, and by the time that final bell rings, I am mentally and physically exhausted. It feels like the control of my outer world is getting weaker and slipping away more each day.

Lynn

**Fantasy-Prone Personality (FPP)**

A fantasy-prone personality is one in which a person experiences a lifelong, extensive involvement in fantasy. A non-pathological FPP is often referred to as having an "overactive imagination" or "living in a dream world."

As a child, I was always daydreaming. My vivid imagination helped me escape from an abusive, unhappy childhood. My parents used to tell me, "You live in your own world." My mother nicknamed me "Sarah Bernhardt" (a French stage actress active in the late nineteenth to early twentieth century). Bernhardt was known for her over-the-top onstage gesturing and highly theatrical lifestyle. I was deeply offended by the nickname and my family's inability or unwillingness to take me seriously. In their view, I overexaggerated and fantasized, and lacked the ability to live in the "real world." In actuality, I saw the reality of my childhood with crystal clarity. I just didn't happen to like it.

Barbara

From as far back as I can remember, my mind seemed to be in adventure mode. One of my first toys was a real microscope. My dad gave me two heavy cast iron scopes. I took them everywhere. I always found things to observe, and saw an entire world underneath that set of lenses. Nature and science were my first loves. As far as I can remember, I did not have any imaginary friends ... well, that were human. My toys and my stuffed animals were my friends. They were alive and had feelings. I had no need for human friends. I would wander the alleys behind my house, looking for horned toad lizards. I loved them with a passion. I would kiss their heads, and we would go on great adventures together. They became my magical beasts! I would take one home with me and create a comfortable bed in a shoe box. I kept it next to me all night. The next morning, I would always set it free. My two dogs, Buster and Sandy, were also my comrades on these great adventures. My mother never knew how far off I would wander into the deserts of Texas. We moved around a lot, but if I wandered far enough, I would find nature, away from people and neighborhoods. Although I eventually became more sociable with my peers, my feet continued to take me into the serene surroundings of the sky, the animals, and all things green. In these places I would experience moments of stillness—ones in which my body escaped me, and I became one with my surroundings.

Lynn

Psychologists Sheryl Wilson and Theodore Barber suggest that anomalous experiencers exhibit FPP (fantasy-prone personalities). Fantasy-proneness is typically regarded as an adaptive response to stress. Wilson and Barber propose that FPPs are atypical and comprise a minority of the populace. As they explain: "There exists a limited group of individuals (possibly 4% of the population) who fantasize a large part of the time, who typically 'see,' 'smell,' 'touch,' and fully experience what they fantasize; and who can be labeled *fantasy-prone personalities.*" [17]

Traits in FPPs include but are not limited to the following:

high hypnotic susceptibility; psychic abilities/ experiences; healing powers; OBEs; vivid or waking dreams; apparitional experiences; and having had an imaginary childhood companion. They report having experienced multiple transpersonal phenomena including ESP, mediumistic trances, automatic writing, and psychic healings. Additionally, they tend to be female.[18]

According to Wilson and Barber, numerous FPPs reported that, "When they were children they believed their dolls had feelings and personalities, while others had imaginary companions, engaged in fantasy games pretending to be 'someone else,' and/or had a childhood belief in guardian angels, spirits, and magical creatures." Wilson and Barber's research determined that the overwhelming majority of FFPs have reported realistic out-of-body experiences, or "weightless floating sensations."[19]

The research of Harvey J. Irwin, Associate Professor of Psychology at the University of New England, Armidale, Australia, demonstrates the strong correlation between fantasy proneness and paranormal beliefs. Irwin contends, "Fantasy proneness may facilitate paranormal belief, and that paranormal belief, in turn, may be conducive to parapsychological experiences."[20] This raises the proverbial question of, "What came first, the chicken or the egg?"

A study conducted by Nicholas Peter Spanos, former Professor of Psychology and Director of the Laboratory for Experimental Hypnosis at Carleton University, achieved similar findings. He concluded that "Among UFO believers, those with stronger propensities toward fantasy production were particularly likely to generate such experiences."[21] Whitley Strieber, horror writer and UAP contact experiencer, is a prime example of an FPP. The late psychologist Robert A. Baker considered Strieber "A classic example of the {fantasy-prone personality} genre. Baker noted that Strieber exhibited such symptoms as being easily hypnotized, having vivid memories and dreams, as well as being a "writer of occult and highly imaginative novels."[22]

Although more in-depth research is required, the aforementioned research findings suggest the existence of a pre-experience, anomalous personality type. It is unfortunate that societal and cultural norms pre-judge and frequently label such persons as freakish, odd, bizarre, and/or peculiar. Possessing the ability to experience alternate realities is a unique quality; however, is this an "abnormal" trait, or merely characteristic of those with extraordinary sensitivities? As word-renowned psychologist and NDE researcher Kenneth Ring, PhD asserts, "There are realms beyond those of consensual reality, and what the majority regards as real is only normative, not definitive. The ASP, on the other hand, tends to perceive beyond normal boundaries, which have hitherto been unacknowledged by science. {These} insights must ultimately be recognized and legitimated."[23]

Psychic people generally notice that they're different from other people in ways that they can't quite define, but it seems to be more than just psychic ability. While it's been shown that most of the world has some level of psychic ability, the distribution is not even. Some people are far more psychic than others. The reasons for this are many and they trace back to a biological predisposition towards psychic ability. People who are very psychic were born that way. The thing to take away from this information is that psychic ability is strongly related to specific physiological characteristics. In other words, it is a genetic predisposition that we can no more alter than our sexual orientation. We feel and act differently because we are wired that way. We simply can't help being who we are. It's not a mental disease, it's a bona fide personality type."[24]

The most interesting people are the unusual. No one writes about or discusses the average, the ordinary, or the common; they write about and discuss the weird, the mad and the different, so if you are one, even though the opinions of others are of no importance, you are, in their eyes, significant to notice and remember.

Donna Lynn Hope, Author

It took me decades to not only accept, but to proudly OWN my uniqueness. I no longer feel like a freak, outcast, or weirdo. I'm fully aware that most people view me as "woo-woo" or "out there." When I'm asked, "In what field do you hold your doctorate?" I answer, "metaphysical science." I typically receive the following responses and/or behaviors: 1. A questioning individual's eyes nearly pop out of their head, while their mouth simultaneously gapes open, nearly to the point of drooling. When the person finally gathers his senses, he manages to stammer ... what, what, what did you say? 2. I get the eye-narrowing, mouth-tightening, angry stare down—with absolutely no commentary, followed by a loud huff of disapproval. 3. The individual squeaks out "oh," turns his back, and walks away as fast as his legs will carry him. I am exaggerating to inject some humor, but not overly so. RARELY, do I hear, "That sounds interesting. Tell me more about it." However, I no longer care what others think. I'm wired differently, and that's fine by me.

As a child, I don't think I even realized there was something "different" about me. I assumed everyone was having out-of-body experiences during the night, flying over the neighborhood and beyond. When I hit my teen years, I realized there was something about me that others would classify as weird. I often knew when people had passed, before I'd been told about it. I sort of felt them leave the world and I often knew under what circumstances. After my near-death experiences things only got stranger. I would meet people and suddenly see events happening in their lives in my mind, kind of like remembering an old movie, or a dream. The intense feelings that came with those visions finally compelled me to share the images with the strangers they concerned. Those revelations came with looks of shock and disbelief. Eventually, I learned to keep these things to myself for fear of being judged, and then came to block them out almost entirely. It's tough to know the tragedies of people's lives and keep them locked away. I know that since my near-death experience I simply AM different. It's like once you've crossed over, you kind of have one foot on the other side to some extent, and it's difficult to shut off. I now consider it a gift and am no longer fearful of the judgment that sometimes comes with it. I am judicious about who I share my gift with.

Penny

Even from an early age, I was an overly sensitive child. I did not relate to my peers so I would explore nature, taking many adventures. As I grew into puberty I tried harder to fit in and was usually bullied. I tried to make friends by giving belongings or my food from my packed lunches away. I came to realize I was different. My heart belonged to nature and the stars in the sky. I grew to hide this part of me, always wearing a mask to hide my true self, in order to "make it" in this world. I am perplexed by the behavior of humanity. I have difficulty understanding why we want to hurt each other, why we live in fear and judgment. Why we are so destructive. The more I accept my inter-dimensional nature, the more I feel alienated from humanity. The most valuable thing I've realized is knowing that I am not alone in this—knowing that there are people like me. I go to work and hide who I truly am, but it's all good, knowing that I have my own "tribe" of like-minded people who share and listen. Let your weird light shine bright so the other weirdos know where to find you.

Lynn

We hope those who consider themselves "weird outsiders" may now reframe their self-perception. Yes, we do have unique personality traits. However, instead of labeling ourselves in a negative manner, let's redefine ourselves positively as curious, unique, creative, and, quite possibly, at the forefront of a new paradigm shift. However, commonalities between experiencers go far beyond personality type. In the next chapter, we will explore our personal extraordinary experiences and subsequent realization that such phenomena is non-local in nature.

# Chapter 3
## Am I Inside or Out of my Mind?

Near-death experiencers, out-of-body experiencers, UAP-related contactees, and those individuals who remember past lives have transcended the time/space/sensory constraints of our 3D realm, opening them to an awareness of the fourth dimension.

Barbara

We are a creation of genetics, environment, neurobiology, and experience. Yet, it is our experiences, and how we interpret, process, and incorporate them into our lives, which shape and define us. Certain individuals have atypical, extraordinary experiences. They are taken on wondrous journeys, traveling outside of everyday, three-dimensional reality, to realms beyond time and space. The authors share their personal experiences with otherworldly phenomena, and how these have reshaped their previous understanding of consciousness.

**Personal Experience of Penny Wilson, RN**
**NDE**
**August 2014**
**Berea, Kentucky**

The Near-Death Experience Research Foundation defines a NDE as: "A lucid experience associated with perceived consciousness apart from the body, and occurring at the time of actual or threatened imminent death."[1]

In August 2014, my life was forever changed by a condition called idiopathic anaphylaxis. The day began like any other. Sitting in the living room with my daughter, sipping a smoothie, I started having trouble swallowing. I went to the kitchen for a glass of water, unsure of what was happening. Within minutes, my chest tightened and breathing grew difficult. I was having a potentially deadly allergic reaction called anaphylaxis.

Years ago, I had been diagnosed with a shellfish allergy. How could this be happening? I wondered. I hadn't been exposed to shellfish. I recognized my situation was life-threatening. I grabbed my epi-pen and injected the drug into my thigh. My symptoms seemed to improve, but not to the extent I'd expected. Years of work as a critical care nurse had taught me the importance of quick action in situations like mine.

My son drove me directly to the emergency room, where I was admitted for treatment. Soon, my shortness of breath returned, and I began having more and more difficulty swallowing. Despite multiple injections of epinephrine, Benadryl, and steroids, my condition continued to deteriorate, until I collapsed and stopped breathing.

It was a strange sort of sensation. One minute I was struggling for breath and for life, and the next I was peaceful and floating ... up, up, up, and out of my body. I briefly saw the hustle and bustle around the person who was lying in the bed below, but I wasn't quite sure who she was. A doctor put a tube in her mouth and down her throat. A respiratory therapist hooked a bag to it and began squeezing air into the lifeless woman's body. Nurses started more IVs. Everyone seemed to be in a state of concerned hurry—but none of it upset me. I was without a care as I watched the scene unfold beneath me. I had no idea my physical life was slipping away.

The next thing I recall was "materializing" in the backseat of my sister's car. She was on her way somewhere, and I got the sense that something was wrong. It was pouring rain, and I worried about her traveling in such treacherous conditions. She pulled into a gas station, beneath the canopy. I watched as she pulled her phone from her bag and opened her Facebook app. I saw her status: "Hang on kiddo, I'm coming." (It was later verified that she had posted this on Facebook while I was lying in a coma in Kentucky.)

Where was she going? Had something happened with the kids? I felt anxious and confused. I tried to speak, but no words came out. I slid my hand down my body and noticed it felt odd, as if it were less dense than normal. With that, everything went dark, and I was instantaneously transported to the other side ... the spirit side.

I spent some time alone in a dark and endless void. Then, a bold spirit came to me. Her hair, a brilliant orange-red, was so bright it looked like she was crowned with fire. She was striking and larger-than-life. It took a moment before I realized the spirit before me was my maternal grandmother. I wept as she held and comforted me. I had so many questions, and as I thought them, the answers came telepathically, from her spirit to mine.

Am I dead?

My grandmother responded:

"No, you aren't dead. There is no death, except that the body becomes useless and is cast away. You've heard it said that energy isn't created or destroyed, it just changes form. This is true in the earthly realm and in the spirit realm. You haven't died, dear one, you've simply changed form."

What fascinating information! I had hoped I would continue to exist in some form after my body died, but I was never completely certain of it. I was amazed to find that my consciousness endured without aid of my body or brain.

I met light beings who I felt innately connected to. I felt linked to every spirit that had ever been created, both on the spirit side and in the earthly realm. I knew each of them, and felt their joys, hopes, and sorrows.

It became clear that everything I had ever thought, said, or done had impacted every other spirit in the universe. I felt shame for the critical judgments I'd placed on others and experienced firsthand how it had harmed them and diminished their energy. I also saw the ripples of love that were created by every kind word and intention I'd ever expressed. My mind and spirit felt as if they might burst with newfound understanding.

At once, I was taken up into a brilliant light, immeasurably brighter than the light emanating from my grandmother. A profoundly deep, energetic vibration shook me to the core. A booming voice said **"I AM"** and I knew immediately who was with me. The spirit of God surrounded me. His light permeated and filled me to overflowing.

As His radiant love soaked into me, all of my questions were answered. Every unknown and unrealized bit of knowledge was brought to light with knowledge from this higher source. Answers from the universe were downloaded into my DNA. I didn't have to try to remember or retain the information, it became part of me. As questions came to mind, the answers were immediately there. All my past hurts were reframed by the bright shining love of God that surrounded me.

His Spirit loved me completely and unconditionally, while my life replayed like a movie on an unseen screen. Watching it, I grieved over things I had done and things I'd neglected to do. I wept for the pain I'd caused others and the pain I had brought on myself. As each scene played, my heart was emptied of all fear and grief by the wonderful love of God. The Supreme Intelligence and Creator of all loved me. Grasping the completeness of His love proved impossible. How could He love me knowing how pitifully I had lived my life? But He didn't judge me. He was merciful and compassionate, overwhelming me with fatherly love. After reviewing my life, God took me on a journey through my DNA, traveling with me through the strands. Each spiral twirled around and over us as we plunged deeper and deeper into my origin. We approached the spark of my existence and I saw that I had originated from God. A simple thought from Him had created me.

God had been with me all along, I just hadn't remembered it until that moment. How had I forgotten such important information? Where had I gotten the idea that He had ever walked away from me? At that moment, I understood I couldn't go any further. I needed to return to my body. I was torn, wanting to stay with my Creator, yet knowing I had more to do on Earth. A powerful force sucked me backward and with a "pop," I woke in my body, the memory of my time with God stored deeply within.

The most impactful lesson of my glorious journey is this: I have come to the realization that our essence, or soul, continues after physical death, in an infinitely more expansive and "real" way.

Penny

**Out-of-Body Experiences**

Researcher Charles T. Tart defines an OBE as: "An experience of mind being elsewhere than body while assessing one's state at the time, as being as clear as ordinary consciousness, not a dream or delusion."[2]

*Image 3-1*
*Illustration of Out-Of-Body Experience*

## Personal Experience of Lynn Miller
## July 2010
## Controlled Out-of-body Experience
## Dyer, Tennessee

I've had spontaneous out-of-body experiences my entire life. Yet, for many years, I did not comprehend what was happening to me. I thought I was having strange, vivid, or lucid dreams in which I remained aware and in total control. I'd heard of OBEs and wondered if they explained my other-worldly adventures. However, my personal paradigm didn't allow for such unbelievable and anomalous experiences.

What concerned me most were the preludes to these so-called lucid dreams. I experienced intense vibrations and loud roaring, similar to a jet engine. I was committed to discovering the cause of these occurrences. I began to research the topic, stumbling upon a podcast featuring William Buhlman (expert on out-of- body experiences). He described, to the finest detail, everything I had experienced.

The realization that I was having out-of-body experiences was life-changing. Even more exciting was learning that, with practice, I could control them. This began my new journey, a journey into consciousness. In July of 2010, I prepared to have my first intentional OBE.

I decided to use the "awake, back to bed" method to initiate a controlled OBE. Utilizing this technique, I set an alarm to awaken myself approximately three to five hours after bedtime. Controlled OBEs are most successfully induced in the early morning hours. In preparation, I awoke at 2 a.m. I remained awake until approximately 3 a.m., reading *Adventures Beyond the Body* by William Buhlman. I went back to bed and turned on a meditative MP3 track to prepare my mind and body. As the vibrations came, I surrendered myself to them, all the while maintaining complete awareness. I thought "Oh my God! It's really happening!" Pushing down my excitement, I calmed myself so as not to cause an abrupt end to the experience. I let the vibrations and loud sounds come, welcoming them. As the resonance stilled, I rose out of my body, out of the bed. I looked around, astonished, then turned and saw my sleeping body. Ecstatic, I reached down and tried to touch the physical "me" that lay in the bed, but my hand went right through it. There was merely a slight resistance. It felt electric.

"This is really happening!" I thought, and then laughed. I was euphoric, almost hysterical, and in an instant, I popped back into my body. My eyes opened, and I laughed and cried with joy! My world had been rocked and would never be the same again.

OBEs have helped me realize our consciousness exists outside of our brains. What an amazing expansion of awareness this has become, to know that life continues beyond this 3D reality we call home. We are products of our culture, failing to see beyond our physicality. Our bodies are merely a temporary vessel for us to learn and experience a corporeal reality.

Consciousness is non-local in nature, and this realization expands our comprehension of the world and people around us. An out-of-body experience is a shift into other dimensions, times, and experiences with non-human intelligent entities. It's a peek into events that are parallel to our existing reality. It is learning how to release the ego and other fear-based paradigms, ingrained in us since birth.

**Contact Experiences**

According to Ted Row, "An Unidentified Aerial Phenomena, UAP, is the visual stimulus that provokes a sighting report of an object or light seen in the sky, the appearance and/or flight dynamics of which do not suggest a logical, conventional flying object and which remains unidentified after close scrutiny of all available evidence by persons who are technically capable of making both a technical identification as well as a common sense identification, if one is possible."[3]

## Personal experience of Barbara Mango
## UAP Sighting: March 20, 1966 12:00 p.m.
## Location: Shawnee, Kansas

On Sunday, March 20, 1966, a blistering heat wave held Shawnee, Kansas, in its grip. For ten consecutive days, record-setting temperatures soared into the eighties. Central air conditioners blasted, while parents dug shorts and T-shirts from summer-clothes bins. Sunday's forecast called for a beautiful, sunny day, with a predicted high of eighty-one degrees.

In my house, nice weather was synonymous with two dreaded words: yard work. My parents ordered me outdoors to rake leaves. I was dutifully raking (and sweating profusely) when, out of the blue, a dark shadow covered our entire lawn. Since when do enormous shadows appear in cloudless blue skies? What was going on? I abruptly dropped my rake and looked upward.

A massive, cylindrical disc hovered five hundred feet above me. It was majestic. I stared up at the matte-silver craft in amazement. It was approximately the circumference of two football fields, with evenly-spaced, oval windows encircling it. I stood in awe of the object as it silently hovered, motionless in the sky.

A low-flying UAP had appeared instantly and in broad daylight, making it impossible to

53

mistake for anything else. Fearless and with utter calm, I gazed up at the craft, while its occupants engaged me in a telepathic conversation.

"FINALLY!" I said. "I've been waiting for you to appear my whole life!"

Silently, the UAP occupants relayed to me:

"We know. You've always understood that the universe is unimaginably vast, containing billions of habitable planets and other intelligent life forms. We ask that you share our mission of peace and concern with the ecology of your planet. We are here to observe you, just as someday, in the near future, you humans will examine moon rocks and lunar soil samples. We ask that you share our mission of peace and concern with the ecology of your planet."

Little did I know, that in a few short years (between 1969 and 1972), 842 pounds of lunar rocks, including core samples, pebbles, sand and dust, would be retrieved from the lunar surface, via six Apollo missions.[4]

Instantaneously, the craft evaporated into thin air. I recall excitedly thinking, "I couldn't hear the UFO because it moved faster than the speed of sound, and I couldn't see it appear or disappear because it traveled faster than the speed of light!" In retrospect, it would have been impossible for me to grasp the laws of energy, waves, vibration, or motion. I was ten years old, in fourth grade. My biggest "scientific challenge" at the time was tackling "health and fundamental science." It definitely wasn't rocket science. Yet, I understood the principles of physics effortlessly. It never occurred to me how I knew—I just KNEW. On a recent whim, I browsed through my old report cards, curious to discover when I had first learned about basic physics. Sixth grade—no, seventh grade—nope, eighth grade—yes. It was when I was fourteen years old, four full years after my sighting.

In this encounter, I received future information regarding lunar samples, which was later publicly verified. I realized that non-human intelligent beings were capable of downloading future scientific data, information which lies beyond time, space, and the human scope of "knowing."

I never doubted the validity of the UAP phenomena I witnessed. The incident was briefly reported in our local paper. However, I wanted hard evidence. I searched for years to no avail. Decades later, I found the confirmation I had long sought in Project Blue Book (PBB). PBB contains declassified Air Force investigations analyzing UFO-related cases/data. Combing for days through the voluminous archives, I was *finally* able to find verification of my childhood UAP sighting (Nara-PBB1-364, Roll 1, March 1–22, 1966 sightings).

**Past-life Regression/Recall**

Past-life therapy is a technique used to regress an individual to recall past lives in order to heal and resolve ailments and situations from the current life. It is based on the ancient belief of reincarnation. The foundation of Hindu philosophy, reincarnation dates back thousands of years. It is the notion that the soul is eternal and incarnates again and again, retaining all the knowledge of events that occur during each lifetime.[5]

Barbara discusses personal past-life experiences and how these have impacted her current incarnation below:

> Each of my past-life memories have taught me valuable and often profound lessons. Yes, several have produced spontaneous physical and phobic healings. Others, however, have been spiritually and emotionally transformational. I'm talking life-altering transformation on a soul/ energetic/DNA level. It's so difficult to put into words. Human vocabulary is utterly limiting in describing transcendent phenomena which defies human perception of space/time. I recall at least two lifetimes in which I was callous, judgmental, indifferent, and yes, even cruel towards others. The memories of my heartless behavior have deeply affected me. In my current incarnation, I feel physical ill when a violent scene occurs on a television show. I become highly anxious if I see an individual being treated unkindly, and am obsessed with justice.
>
> Perhaps my abhorrence to violence is rooted in my past life as a highly esteemed Roman soldier. Recalling this life appalls me. I have always regarded the Roman Empire as an imperialist society who glorified violence. Gladiatorial combatants entertained blood-thirsty audiences, battling fellow gladiators, wild animals, and condemned criminals. Roman emperors were, by and large, infamous for tyrannical reigns of often perverse violence. Just thinking about this fierce, often sadistic

empire sickens me. I always believed (or wanted to believe) that I had lived a glorious life in ancient Greece, a society unparalleled for its philosophical, artistic, scientific, political, and metaphysical contributions to Western society. A peaceful civilization, home to numerous great and historically unforgettable names: Hippocrates, Euclid, Pythagoras, Plato, Aristotle, and Socrates. Thus, imagine my repulsion when I spontaneously recalled a life in ancient Rome, described below.

I am a powerfully built soldier in the prime of my life, handsome, well-educated, and possessed with an innate, commanding presence. I am standing on the stairs of a sacred temple, overlooking a forum, an ancient type of civic center. A throng of perhaps 1,000 individuals are wildly cheering my triumphant, victorious return from Caesar's Civil War. I am dressed in a white tunic. A red paludamentum (a cloak or cape fastened at one shoulder, worn by military commanders) is draped around my shoulders. I look down at my feet, covered in high-quality leather sandals. It is a warm, cloudless day. However, I do not feel like a hero. I feel like an imposter. I am, by nature, a kind, fair, and insightful man. Yet, I had allowed culture and familial expectations to push me into a life of violence. How had this happened? I was sickened by the knowledge that as a high-ranking soldier, I had ordered the killing of countless human beings. I felt conflicted, disgusted with myself, and immeasurably sad. As suddenly as it began, that life faded, and I was once again in my current incarnation.

I am lucky these past-life memories occurred when I was a child, so that my worldview was transformed early in life. I've always realized that as interconnected beings, we must treat one another with kindness, respect, and without judgment.

Jim Tucker, MD, current researcher and an associate psychiatry professor at the UVA Medical Center's Division of Perceptual Studies, believes the theory of non-locality (non-local consciousness) may explain how memories of one person might transfer to another lifetime. Tucker asserts: "The discovery of quantum physics indicates the physical world is affected by, and even derived from the non-physical, from consciousness. If that's true, then consciousness doesn't require a three-pound brain to exist, and so there's no reason to think that consciousness would end with it. It's conceivable that in some way consciousness could be expressed in a new life." [6]

The aforementioned paranormal contact experiences support the theory that we live in a multidimensional reality. Physicists and scientists refer to this as the Quantum Hologram (QH) Theory of Consciousness. QH theory maintains that humans are both physical and spiritual beings. Yet, mainstream science tenaciously clings to traditional paradigms, while continuing to dismiss the validity of experiential claims. Why is this? Two main factors are at play.

First, nearly all government funding is awarded to traditional science. According to philosopher Ilja Maso, "Most current researchers employ the materialist hypothesis because it "attracts" most of the funding, achieves the most striking results, and is thought to employ the brightest minds. The more a vision deviates from this materialist paradigm, the lower its status and the less money it receives." Secondly, materialists tend to reject any challenge to their rigidly held model.[7]

The late philosopher Thomas Kuhn stated that the non-materialist theory is often incongruent with the traditional model. As he explained: "All research results that cannot be accounted for by the prevailing worldview are labeled 'anomalies' because they threaten the existing paradigm and challenge the expectations raised by this paradigm. Needless to say, such anomalies are initially overlooked, ignored, rejected as aberrations, or even ridiculed."[8]

Let's now take a closer look at the debate between science and experiencers of NDEs, OBEs, UAP-related contact, and past-life recall. How do skeptics view these phenomena, and why are experiencers insisting they are non-local in nature?

# Chapter 4
## Divided Minds
## Skeptic vs. Experiencer

*My fundamental premise about the brain is that its
workings—what we sometimes call "mind"—are a
consequence of its anatomy and physiology and nothing
more.[1]*
Astronomer and popular science writer
Carl Sagan

*I am a scientist. I think the way to the truth is by
investigation. I suspect that telepathy, clairvoyance,
psychokinesis, and life after death do not exist because I have
been looking in vain for them for 25 years.[2]*
Psychologist and researcher
Susan Blackmore

*If what you mean by "soul" is something immaterial and
immortal, something that exists independently of the brain,
then souls do not exist. This is old hat for most psychologists
and philosophers, the stuff of introductory lectures.[3]*
Psychologist Paul Bloom,
author of Descartes Baby

*Consciousness is nothing more than "by-products of the
brain's electrical and chemical processes.[4]*
Authors Mario Beauregard
and Denyse O'Leary

In 2005, the journal *Science* published an issue listing the
top 125 questions scientists have yet to answer. The most
significant inquiry was, "What is the universe made of?"
immediately followed by, "What is the biological basis of

consciousness?" This question may be reframed by asking, "Does consciousness have a biological basis at all?"[5] Herein lies the crux of the debate between science and medicine concerning NDEs, OBEs, UAP-related contact, and PLR. Why does mainstream science cling so tenaciously to the materialist paradigm, and continue to ridicule experiencers?

Research on the pro/con arguments pertaining to NDEs/OBEs/UAP-related contact, and past-life memories is exhaustive. Entire books are devoted to these subjects. Thus, our discussion is limited to a brief debate concerning the most commonly challenged evidence. Additionally, we present personal arguments from our unique experiencer perspectives.

*Image 4-1*
*Illustration of Skepticism*

### NDEs

Historically, a NDE was considered to have occurred when a person was declared clinically dead. However, studies now show that patients who are critically ill, undergoing surgery, or have a near-brush with death can also have near-death experiences. Clinical death is determined by the following factors: cessation of heartbeat, known as asystole; the absence of electrical activity in the cortex of the brain; lack of brainstem function; fixed and dilated pupils; and absence of a gag reflex. Brain death begins with unconsciousness, which typically occurs within ten to thirty seconds of cessation of heartbeat. If the heart is not immediately restarted, a complete cessation of electrical activity in the cerebral cortex inevitably occurs, resulting in a flat EEG within minutes. Thus, measurable electrical brain activity, responsible for conscious thought, ceases to exist. At this point, the brain stem loses complete function and "dies" within fifteen to sixty minutes. Pim van Lommel compares the brain in this state to "A computer that has been disconnected from its power supply, unplugged, and all its circuits disabled."[6]

Yet, this is precisely the moment when NDEers report having rich and incredibly lucid experiences, explaining them as more real than real, indescribable, and beauteous. No two NDEs are identical; however, there are many core elements: out-of-body experiences; heightened lucidity and sensory input; entering/passing through a tunnel; seeing a brilliant light; meeting deceased loved ones or spiritual beings; telepathic communication; time and space losing all meaning; a life review; receiving downloads of previously unknown information; traveling to other dimensions; and lastly, reaching a non-crossable border where a voluntary or involuntary return to the physical body occurs.

Traditional science, however, continues to attribute NDEs to physiological, psychological, and pharmacological factors. These include but are not limited to temporal lobe seizures, REM sleep disturbance, dissociation, insufficient anesthesia, hallucinations, and anoxia (a severe, life-threatening oxygen deficiency).

I find the oxygen deprivation (anoxia)

theory as an explanation for NDEs to be conjecture, at best. My first near-death experience began during an anaphylactic attack and respiratory failure; however, it continued after I was placed on a mechanical ventilator and properly oxygenated. During my comatose state, I remained out of body, hovering between my hospital room and another dimension where I was in a dark void, and then progressed on to a bright white light. During my out-of-body state, I saw what was going on in my hospital room, the ICU waiting room, and my best friend's car as she traveled from Wisconsin to Kentucky.

I related to my family what I saw while out-of-body and did so with stunning accuracy. This does not speak to an oxygen-deprived state, as my thoughts were cohesive, organized, and consistently accurate to the finest detail. When I awoke from my coma, I told my friend that I had seen her traveling in her car and asked why she was wearing clothes that didn't match. She looked shocked and explained that she had just put on the clothes from the laundry that had been folded at the end of her bed. I told her I saw what she'd posted on Facebook as she traveled to see me: "Hang on kiddo, I'm coming." There was no way I could have known about that post. I also told my son that I had heard his conversation with my friend about my not having a living will. When my daughter arrived to see me, I told her that I'd seen her in my room in her red plaid shirt with her sleeves rolled up. She confirmed she'd worn that shirt to visit me a few days earlier while I was still in a coma.

Penny

Materialist science argues that complex states of consciousness described by experiencers (including vivid mentation, sensory perception, and memory) are impossible during clinical death. It asserts that since NDEs cannot be *irrefutably* proven, the non-materialist paradigm is inherently flawed. Educator and philosopher Neal Grossman emphasizes the extent to which traditional science rejects the non-materialist model by recalling a dialogue amongst colleagues: "One conversation in particular caused me to see more clearly the fundamental irrationality of academics with respect to the evidence against materialism ... I asked, "What will it take short of having a near-death experience yourself to convince you it's real? Very nonchalantly, without batting an eye, the response was: "Even if I were to have a near-death experience myself, I would conclude that I was hallucinating, rather than believe my mind can exist independently of my brain."[7]

Yet, experiencers consistently describe NDEs as unbelievably lucid, life-transforming, and *realer than real*. Prior to his own NDE, academic neurosurgeon and best-selling author Eben Alexander upheld the traditional theory. Afterward, his perspective changed radically. As he explains:

> Like many other scientific skeptics, I refused to even review the data relevant to the questions concerning these phenomena. I prejudged the data, and those providing it because my limited perspective failed to provide the foggiest notion how such things might actually happen. Those who assert that there is no evidence for phenomena indicative of extended consciousness, in spite of overwhelming evidence to the contrary, are willfully ignorant. They believe they know the

truth without needing to look at the facts.[8]

### NDE-related OBEs

Although OBEs, in and of themselves, may be studied as a singular contact modality, they are also considered a subcategory of NDEs. Virtually all NDEers report experiencing a separation of the soul from their physical body, while simultaneously observing it from "a perspective different from the body's actual location." Most often they describe floating above their physical bodies; observing both themselves and surrounding activity in a detached, calm manner.

Most non-materialist researchers are in agreement that the OBE, or separation of consciousness from the physical body, is one of the most scientifically verifiable aspects of a NDE. Near-death researcher Kenneth Ring further substantiates the detailed perception of such NDE-related OBEs, by stating: "The perceptions are just too fine-grained in their details and too telling in their appropriateness—they are just the kind of thing one would expect to be reported if individuals were really able to see with extraordinary clarity from an elevated position near the ceiling—to be glibly written off on the grounds that they are simply not possible."[9]

While out-of-body, NDE vision is vastly different from normal visual perception. Nearly all case subjects report having *greater* than 360-degree vision, or omnidirectional awareness. NDEers often report spherical, three-dimensional visual awareness simultaneously in all directions—forward, backward, right, left, above, and below. The most compelling cases of omnidirectional awareness are those reported in the congenitally blind. Psychologists Kenneth Ring and Sharon Cooper conducted the most in-depth study ever undertaken of NDEs in the blind. The objective of their study was to determine if the blind experience the same veridical occurrences as sighted persons. Thirty-one subjects were chosen for the study. Of these, fourteen were congenitally blind. Research ascertained that narratives of the blind were *indistinguishable* from those of the sighted.[10]

Vicki Noratuk is congenitally blind. After a near-fatal car accident and suffering from brain damage, Vicki was rushed to the hospital in a coma. She recalls her experience by stating, "And it was frightening because I'm not accustomed to see things visually, because I never had before. And initially it was pretty scary."[11]

Upon resuscitation, Vicki described seeing her crumpled Volkswagen van. Additionally, she "saw" herself floating above the stretcher and traveling to the hospital's roof, where she experienced a 360-degree panoramic view of the hospital grounds. Vicki's surgical team later verified her accurate description and precise account of hospital grounds. In a later interview, Vicki was asked to compare her dreams and NDE. When asked if she had ever experienced visual perception while dreaming, she responded, "Nothing. No color, no sight of any sort, no shadows, no light, no nothing."[12]

Ring and Cooper describe the cases of two *sighted* NDEers. Their descriptions are also remarkable:

> I will never forget the look on the surgeon's face when I told him that I went through the OBE phenomenon during the operation. (Afterwards) I then asked the surgeon whether he was sitting on a green stool with a white top on it. He replied yes. He then said, "But you could not have seen that from where you were lying on the operating table." I then said to him that I did not see that from where I was lying, but that I had seen it from where I was detached from my body looking down from above during this NDE phenomenon. This remark caused an even stranger look on his face.[13]

> I could see everything. And I do mean

everything! I could see the top of the light on the ceiling, and the underside of the stretcher. I could see the tiles on the ceiling and the tiles on the floor, simultaneously. Three hundred sixty-degree spherical vision. And not just spherical. Detailed! I could see every single hair and the follicle out of which it grew on the head of the nurse standing beside the stretcher. At the time I knew exactly how many hairs there were to look at.[14]

## OBEs

It was just over thirty years ago that I had the dramatic out-of-body experience that convinced me of the reality of psychic phenomena and launched me on a crusade to show those closed-minded scientists that consciousness could reach beyond the body and that death was not the end. Just a few years of careful experiments changed all that. I found no psychic phenomena—only wishful thinking, self-deception, experimental error and, occasionally, fraud. I became a skeptic.[15]

Susan Blackmore, Psychologist


If I ever had an out-of-body experience, I'd try to come back to a different one.[16]

Tom Wilson , Actor, Writer, Comedian


Though astral projection practitioners insist their experiences are real, their evidence is all anecdotal—just as a person who takes peyote or LSD may be truly convinced that they interacted with God, dead people, or angels while in their altered state. Astral projection is an entertaining and harmless pastime that can seem profound, and in some cases even life changing. But there's no evidence that out-of-body-experiences happen outside the body instead of inside the brain. Until the existence of an astral plane can be proven—and made accessible—there's always the continuing adventures of the Sorcerer Supreme.[17]

Benjamin Radford

Why does traditional science continue to reject the validity of OBEs? Science requires "hard evidence." Yet, how does science attempt to measure how, when, or even if a person's consciousness enters or leaves its body? Due to the seeming insurmountability of this task, mainstream science rejects the very notion of OBEs.[18]

In co-author Lynn Miller's view, OBE skeptics fall into two categories: total disbelievers, and/or those who know just enough about them to view them as scary or demonic.

She explains her firsthand experiences with both skeptics and those who misinterpret the OBE phenomena:

### Skeptics

I once confided to a friend about OBEs I had with my guides and with non-human entities. I was told that I was deranged and had delusions of grandeur—that I must be really miserable with my life to fabricate such fantasies. This person believed that I was unstable and out of touch with reality. I was pitied.

I had a boyfriend that insisted that I was

68

experiencing some type of dissociative disorder. Like me, he holds a degree in psychology. I tried to explain to him the difference between an OBE experience and dissociative episode. Trying to inject some humor, he callously responded, "Well then, the next time you are having an out-of-body experience, why don't you clean your house!"

Numerous individuals think they understand OBEs, yet instead, completely misinterpret them. I find these people to be worse than complete non-believers because they damage the reputation, idea, beauty, and magnificence of the experience. They are full of false information. Unfortunately, the movie industry is guilty of portraying terrifying OBEs. The movie franchise *Insidious* deals with the potential horror of astral projection. The viewer is led to believe he may become demonically possessed upon entering the astral plane. I am a member of several astral projection forums and groups and have read thousands of experiencer accounts. The false assumption (that one may be possessed during astral travel) is quite common. Other false convictions include being attacked by negative entities, remaining powerless to return to the physical body, and in extreme cases, dying.

Many religious beliefs prohibit the practice of OBEs. Several years ago, I was reading *Adventures in the Afterlife*, by William Buhlman. Evidently, my sanctimonious brother did not approve of my reading material. I was just finishing a chapter, when BAM, he hurled his Bible at me, narrowly missing my head. He told me, in no uncertain terms, "You are committing blasphemy against God. The Bible states that we are not allowed to know about the afterlife; thus, you are going to Hell. You need to start reading the Bible."

I felt intimidated and beaten down. I could normally handle most of his abusive, rage-filled behavior. But this? This hurt me to the very core. My OBEs are so personal, beautiful, and empowering. Yet, his reaction was vile. I was also infuriated. My brother was representing the worst aspects of religious fervor: judgment and ridicule.

Lynn

### Misinterpretation/Dreaming

Are OBEs merely just dreams, as many people assume they are? "Dreams are the stories the brain tells during sleep—they're a collection of clips, images, feelings, and memories that involuntarily occur during the REM (rapid eye movement) stage of slumber."[19]

I am frequently told that my OBEs are dreams. I strongly disagree with this. They are not dreams as most individuals understand them. Every night our minds experience non-locality of consciousness, or, what I call "adventures in consciousness." The difference between dreams and OBEs is the heightened awareness and control that one experiences during OBEs. Dreams are clips and stories; they are a means in which we deal with daily life. Most make extraordinarily little sense at all, though they do portray a deeper meaning

70

that often reflects our anxiety and inner fears. Dreams are often full of symbolic representations of important things in our lives, and they play an especially key role in our mental health.

There is a spectrum of consciousness that spans across dreams, lucid dreams, and OBEs. In dreams we have no awareness or self-control, it just plays out. In lucid dreams, the dreamer may be aware that he is dreaming. The OBE is a complete separation of consciousness from our body where awareness is complete; this awareness is so acute that it feels as if we are in the conscious waking state.

One cannot assume that OBEs are dreams, especially if they have never experienced one.

Lynn

Neurologist Dr. Olaf Blanke, a skeptic of out-of-body experiences, conducted an experiment at the University Hospitals of Geneva and Lausanne in Switzerland. The experiment's goal was to provide a scientific explanation for this paranormal phenomenon. It attempted to discredit not only those who believe the mind and spirit are separate from the brain, but also sought to debunk contact experiences and other paranormal claims.[20]

Out-of-body experiences (OBEs) are curious, usually brief sensations in which a person's consciousness seems to become detached from the body and take up a remote viewing position. Here we describe the repeated induction of this experience by focal electrical stimulation of the brain's right angular gyrus in a patient who was undergoing

71

evaluation for epilepsy treatment. Stimulation at this site also elicited illusory transformations of the patient's arm and legs (complex somatosensory responses) and whole-body displacements (vestibular responses), indicating that out-of-body experiences may reflect a failure by the brain to integrate complex somatosensory and vestibular information.[21]

Blanke and the Swiss researchers mapped the brain activity of a 43-year-old woman suffering from complex partial seizures by implanting electrodes to stimulate portions of her right temporal lobe. The specific site that induced OBEs in the right temporal lobe was the angular gyrus. In the initial simulations, she reported "sinking into the bed" or "falling from a height." After the current was increased, she reported seeing herself lying in bed from above; however, she only saw her legs and lower trunk. Other parts of the right temporal lobe through focal electrical stimulation also induced motor, somatosensory (touch, pressure, temperature), and auditory responses. Blanke concluded that failure of this part of the brain to integrate properly results in the experience of dissociation of self from the body; hence, the OBE sensation.[22]

Lynn Miller considers the findings of Blanke and other neurological researchers fascinating. As she explains:

I have a background in both psychology and biology. I find the physiology of the brain very intriguing. From neurons to neurotransmitters, the brain, to a large degree, remains a mystery. I support the data of Blanke, most especially, data concerning subjects who experienced "dis-association of self" from the physical body. Let us look at the definition of disassociation in relation to mental disorders.

72

Dissociative disorders involve problems with memory, identity, emotion, perception, behavior, and sense of self. Dissociative symptoms can potentially disrupt every area of mental functioning.

Examples of dissociative symptoms include the experience of detachment or feeling as if one is outside one's body, and loss of memory or amnesia. Dissociative disorders are frequently associated with a previous experience of trauma.[23]

As Lynn explains:

> I disagree with Blanke's conclusion that OBEers are experiencing dissociative episodes. During an OBE, there is no loss of self at all. One is very much aware of himself. It is often said that OBEers see their bodies from below as they float above. Yes, this can happen. However, it is more of a rarity than the norm. I experienced three decades of spontaneous OBEs before I realized what was happening. Never, in all those years, did I see my body as I floated above it. What I did experience on numerous occasions were the infamous "false awakenings." I rose and left my bed to complete routine activities only to discover that things were "off." For example, the clock displayed incorrect time; items were moved or gone entirely, only to be replaced by different objects. If my OBE occurred at night, my room glowed. Most frequently, it would appear bluish in color. I experienced adventures such as flying over towns and country sides and communicated both with human and non-human beings.

Live Science contributor Benjamin Radford disbelieves the legitimacy of OBEs. As he states: Because of the lack of scientific evidence to prove the validity of such phenomena, "It is impossible to scientifically measure whether a person's spirit leaves or enters the body. OBEs are rejected by mainstream science because there is no material evidence that consciousness can exist outside the brain."[24]

Traditionally, science has maintained that OBEs can be induced/or explained by brain traumas, sensory deprivation, near-death experiences, dissociative episodes, dreaming, psychedelic drugs, sleep apnea, vestibular disorders, and electrical stimulation. Much discussion among contemporary skeptics has centered around dreaming, brain stimulation, and vestibular disorders.

### Vestibular Disorders

The vestibular system is defined as "a collection of structures in your inner ear that provides you with your sense of balance and an awareness of your spatial orientation, meaning a sense of whether you are right-side-up or upside-down. Your brain then integrates that information with other sensory information from your body to coordinate smooth and well-timed body movements."[25]

A recent study conducted by Lopez and Elziere, entitled "Out-of-Body Experience in Vestibular Disorders - A prospective study of 210 patients with dizziness," was published in the journal *Cortex*. It proposed that individuals suffering from vestibular disorders have a "significantly higher occurrence" of OBEs than the general public. Fourteen percent of patients reported out-of-body experiences. This disorder may cause dizziness, imbalance, problems with spatial orientation, lightheadedness, and floating sensations. One of the patients in this study explains the sensation in this way: "Feeling like I'm outside of myself. I feel like I'm not in myself." Another said it felt like this: "I was divided into two persons, one who had not changed posture, and another new person on the right, looking somewhat outwardly. Then the two somatic individuals approached each other, merged, and the vertigo disappeared."[26]

At the conclusion of their study, Lopez and Elziere stated, "Altogether, our data indicates that OBE in patients with dizziness may arise from a combination of perceptual incoherence evoked by the vestibular dysfunction with psychological factors (depersonalization-derealization, depression, and anxiety) and neurological factors (a migraine)."[27]

Penny Wilson asserts:

> I'm familiar with the sensation Lopez and Elziere refer to. Occasionally, I've had extreme vertigo or near-fainting episodes. In these instances, I feel a loss of control, as if I am being ripped from reality and separated from myself—a sense of depersonalization. However, during my NDE-related OBEs, I have never felt detached from myself emotionally. I always remain exceedingly lucid and in control. I believe it would be beneficial for researchers to interview people who have had multiple experiences of vertigo, fainting, OBEs, NDEs, hallucinations, etc. This would enable them to gather opinions from a subset of the population who are able to compare and contrast each experience.

### UAP-Related Contact

The preferred term for Extra Terrestrials (ETs) is a non-human intelligent being, (NHIB). Like we "earthlings," they have evolved (or presented) in a myriad of shapes, colors, and sizes. Unfortunately, fear of these, or any unknown entities, remains embedded in the psyche of many. Thus, it is significant to understand that NHIBs are not the Martians or little green men of ET folklore. Rather, they are

75

extraordinarily intelligent, perceptive, and highly evolved beings, often with distinct personalities.

Mainstream science is particularly dismissive of UAP-related contact. Astronomer and skeptic Alan Hale has stated:

> I have seen diverse phenomena as: fireballs, rocket launches, ionospheric experiments, satellite re-entries, comets, auroras, bright planets, novae, orbiting satellites, and high-altitude balloons—all of which have been reported as "UFOs" by uninformed witnesses. If indeed there are alien space craft flying around earth with the frequency which UFO devotees are claiming, then I must ask how come I have never seen anything remotely resembling such an object, while at the same time I have managed to see all these various other types of phenomena.[28]

> Eyewitness reports of actual space-ships and actual extraterrestrials are, in themselves, totally unreliable. There have been innumerable eye-witness reports of almost everything that most rational people do not care to accept ... ghosts, angels, levitation, zombies, werewolves, and so on.[29]

> Isaac Asimov

> I'm sure the universe is full of intelligent life. It's just been too intelligent to come here.[30]
>
> Arthur C. Clarke

Numerous astronauts report having personally witnessed UAP. On May 16, 1963, Astronaut Gordon Cooper lifted off for the final solo space flight that NASA launched. His mission was to orbit the Earth twenty-two times. During his final orbit, Cooper reported to the tracking station (Muchea) near Perth,

Australia, seeing a glowing, greenish object ahead of him quickly approaching his capsule.[31] The UAP was detected on Muchea's radar tracking system. Although the sighting was reported by the National Broadcast System, reporters were later forbidden to interview Cooper about the incident. Cooper, a strong believer in UAP-related phenomena, testified before the United Nations, stating: "I believe that these extraterrestrial vehicles and their crews are visiting this planet from other planets ... Most other astronauts were reluctant to discuss UFOs. I did have occasion in 1951 to have two days of observation of many flights of them. Of different sizes, flying in fighter formation, generally from east to west over Europe."[32]

Several astronauts have since spoken openly about UAP sightings:

> I think that Walter Schirra aboard Mercury 8 was the first of the astronauts to use the code name "Santa Claus" to indicate the presence of flying saucers next to space capsules. However, his announcements were barely noticed by the general public. It was a little different when James Lovell on board the Apollo 8 command module came out from behind the moon and said for everybody to hear: "PLEASE BE INFORMED THAT THERE IS A SANTA CLAUS." Even though this happened on Christmas Day 1968, many people sensed a hidden meaning in those words.

> Astronaut Jim Lovell recalled a personal sighting, which occurred during the 1965 Apollo 8 mission:

> Lovell: BOGEY AT 10 O'CLOCK HIGH.
> Capcom: This is Houston. Say again?
> Lovell: SAID WE HAVE A BOGEY AT 10 O'CLOCK HIGH.

> Capcom: Gemini 7, is that the booster or is that an actual sighting?
>
> Lovell: WE HAVE SEVERAL ... ACTUAL SIGHTINGS.
>
> Capcom: ... Estimated distance or size?
>
> Lovell: WE ALSO HAVE THE BOOSTER IN SIGHT ...
>
> - Exchange between Cape Kennedy and Astronaut James Lovell, December 1965

Reflecting on his historic flight on its fiftieth anniversary, Lovell expressed that prior to the mission:

> I thought, my world is only as far as the eye can see. In the country, mountains, hills or a grove of trees can restrict my world. In cities, tall buildings define my world and in this cathedral, our world exists within these walls. But seeing the Earth 240,000 miles away, my world suddenly expanded to infinity. I put my thumb up to the window and completely hid the Earth. Just think, over five billion people, everything I ever knew was behind my thumb. As I observed the Earth, I realized my home is a small planet, one of nine in the Solar System. It is just a mere speck in our Milky Way galaxy and lost to oblivion in the universe.[33]

Years later, Lovell addressed an enthusiastic crowd at Southern Illinois University, Edwardsville. When asked about the possibility of extraterrestrial life, Lovell stated, "We would all be naive to think we're the only intelligent life in the universe."[34]

The late Dr. Edgar Mitchell, who was the sixth man to walk on the moon, publicly expressed his opinion that he was "90 percent sure that many of the thousands of unidentified flying objects, or UFOs, recorded since the 1940s, belong to visitors from other planets." Dateline NBC conducted an interview with Mitchell on April 19, 1996, during which he discussed meeting with officials from three countries who claimed to have had personal encounters with extraterrestrials. He offered his opinion that the "Evidence for such 'alien' contact was 'very strong' and 'classified' by governments, who were covering up visitations and the existence of alien beings' bodies."[35]

In the 1990s, Pulitzer Prize winning psychiatrist Dr. John Mack studied dozens of people who said they had had such contact with aliens, culminating in his book *Abduction: Human Encounters with Aliens* in 1994. In it, he focused less on whether aliens were real than on the spiritual effects of perceived encounters, arguing that "The abduction phenomenon has important philosophical, spiritual and social implications" for everyone.[36]

In 1996, Mack was interviewed by Nova, a PBS television affiliate. Highlights from the interview include:

> NOVA: What do you think of the work of people like Michael Persinger and Robert Baker who have these complicated theories about neurology, or they charge that hypnagogic hallucinations being at the root of these (UFO-contact experiences) perceived— these experiences?
>
> MACK: These experiences often occur in literal consciousness. Not in a hypnagogic or dreamlike state. The person may be in their bedroom quite wide awake. The beings show up. And there they are, and the experience begins. That they're not occurring in any

79

dreamlike state. Now sometimes they do occur when a person is dozing off or in a hypnagogic state. But very frequently not.

Also, any theory that is going to look upon this as a purely endogenous phenomenon, by which I mean generated purely from the psyche of the person themselves. Which is a kind of arrogance too, really. Because it means that we just can't accept the notion there could be another intelligence at work here. Which is a much more economical explanation. But if we must find a theory within ourselves, then we should keep in mind that any theory that's going to even begin to address this, has to consider four factors:

Number one, the extreme consistency of the stories from person after person. Which you would not get simply by stimulating the temporal lobes. You would get very variable idiosyncratic responses that would differ a great deal from person to person.

Number two, you would have to deal with the fact that there is no ordinary experiential basis for this. In other words, there's nothing in their life experience that could have given rise to this, other than what they say. In other words, there's no mental condition that could explain it.

Third, you have to account for the physical aspects: the cuts and the other lesions on their bodies, which do not follow any psycho-dynamic distribution ...

Fourth, the tight association with UFOs, which are often observed in the community, by the media, independent of the person having the abduction experience, who may not have seen the UFO at all, but reads or sees on the television the next day that a UFO passed near where they were when they had an abduction experience.[37]

### Past-Life Recall

There is no such thing as reincarnation. If you have imagined a previous life, that doesn't make it true, no matter how true it may seem to you or how useful you think it is for giving you true insights.[38]

Richard McNally, Clinical Psychologist,
Harvard University

Past-life regression therapy (PLRT) is a hypnotic technique used to recover memories of past lives or incarnations. It is utilized to resolve psychological issues, emotional issues, phobias, pain, and chronic health conditions affecting our current lives. PLRT maintains that traumatic experiences from a past life may "carry over" into a current incarnation. Thus, only by reliving and discovering the *origins* of present-day challenges, may we identify, release, and permanently heal them. Unlike traditional psychotherapy, PLRT often alleviates symptoms in as little as one session.

Yet, mainstream psychiatry has traditionally rejected the concept of past lives, and refuses to accept regression therapy as an approved healing modality. In an article for Live Science magazine, skeptic Melinda Wenner writes, "There are at least two attractive features of past-life regressions. Since therapists charge by the hour, the need to explore centuries instead of years will greatly extend the length of time a patient will need to be 'treated,' thereby increasing the cost of therapy. Secondly, the therapist and patient can usually speculate wildly without much fear of being contradicted by the facts."[39]

81

Traditional researchers contend past-life regression (PLR) is the *alleged* journeying into one's past lives while hypnotized. They maintain that although many patients "recall" past lives, it is highly probable that their memories are "false memories" (fabricated or distorted recollections of an event). Researchers suggest these memories are from experiences in this life, pure products of the imagination, intentional or unintentional suggestions from the hypnotist, or pure confabulations.

Researcher Maarten Peters of Maastricht University in The Netherlands asserts the following: "People who are likely to make these kinds of errors might end up convincing themselves of things that aren't true. When people who are prone to making these mistakes undergo hypnosis and are repeatedly asked to talk about a potential idea—like a past life—they might, as they grow more familiar with it, eventually convert the idea into a full-blown false memory."[40]

Richard McNally, a clinical psychologist at Harvard University, suggests that highly creative people are more prone to imagine past lives. As he explained in a phone interview with NBC News, "People who commonly make source-monitoring errors respond to and imagine experiences more strongly than the average person, and they also tend to be more creative. It might be harder to discriminate between a vivid image that you'd generated yourself and the memory of a perception of something you actually saw." [41]

In an article published in *The Journal of Medical Ethics and History of Medicine,* author Gabriel Andrade argues that past-life regression is medically unethical. Andrade, a Professor of Ethics at Xavier University, Aruba, believes that although PLRT (past-life regression therapy) has been used to treat anxiety, mood disorders, and gender dysphoria, it is not evidence-based therapy. He asserts:

> Past life regression is based on the reincarnation hypothesis, but this hypothesis is not supported by evidence, and in fact, it

faces some insurmountable conceptual problems. If patients are not fully informed about these problems, they cannot provide informed consent, and hence, the principle of autonomy is violated. Second, past life regression therapy has a significant risk of implanting false memories in patients, and thus, causing significant harm. This is a violation of the principle of non-malfeasance, which is surely the most important principle in medical ethics.[42]

The medical field remains wary of PLR. Many considered it a pseudo-science, as medical internships in the field do not exist. Dr. Brian Weiss is a world-renowned psychiatrist, hypnotherapist, and pioneer in the field of past-life regression. His internationally best-selling book, *Many Lives, Many Masters* (1988), presented an innovative method for treating phobias and physical ailments unresponsive to traditional medical treatment. The book details Weiss' work with a patient referred to as "Catherine." While under hypnosis, Catherine recalled multiple past lives, relieving her of paralyzing phobias. However, its publication caused Weiss to be censured by the medical community. The American Psychological Association (1995) released an official statement regarding Weiss' therapeutic method, stating:

The American Psychiatric Association believes that past life regression therapy is pure quackery. As in other areas of medicine, psychiatric diagnosis and treatment today are based on objective scientific evidence. There is no accepted scientific evidence to support the existence of past lives let alone the validity of past life regression therapy.[43]

Although many still consider his methodology pseudo-scientific, Weiss, who has since gained widespread respect in the field of past-life regression therapy, states:

> Doctors are emailing me. They're not so concerned with their reputations and careers. We can talk about this openly. And it's not just psychiatrists, but surgeons and architects.[44]

Another trailblazer in the field of PLRT was the late Dr. Roger Woolger, psychotherapist and co-founder of the European Association for Regression Therapy. In an interview with renowned past life-regressionist Carol Bowman, he stated, "From nearly a decade of taking clients and colleagues through past life experiences and continuing my own personal explorations, I have come to regard this technique as one of the most concentrated and powerful tools available to psychotherapy short of psychedelic drugs."[45]

Cutting-edge research into the fields of Quantum Hypnotherapy (QH) and forensic handwriting analysis continues to investigate the validity of PLRT. The late Dolores Cannon was a renowned past-life hypnotherapist and innovator of Quantum Hypnotherapy. This technique is considered an advanced therapeutic method which combines hypnotherapy and quantum physics. The ultimate goal of QH is to achieve a much deeper state of trance where the physical and mental states are experienced at the vibratory or quantum level. In this quantum state of mind-body, healing happens beyond any conscious effort like suggestions or visualization ... QH is very close to the process of self-realization or achieving a state of awareness.[46]

Forensic handwriting combines the fields of optics, psychology, and physiology. Its analysis examines letter formation; spacing, slant, and letter size; unusual formation/flourishes; pen lifts; shading and pressure; spelling, grammar, punctuation, and phrasing. With its exhaustive list of criteria, it is no wonder that forensic documentation is utilized by the FBI and CIA. Additionally, the majority of courts uphold that forensic document examination is admissible evidence, based on the supposition that no two people's handwriting can be identical.

Forensic Scientist Vikram Raj Singh Chauhan has done extensive research on the correlation between past lives and forensic handwriting analysis. His findings have been presented at the National Conference of Forensic Scientists at Bundelkhand University, Jhansi, India. According to Chauhan, a six-year-old boy named Taranjit Singh from the village of Alluna Miana, India, has claimed since he was two that he had previously lived as Satnam Singh. Taranjit insisted that Satnam had lived in the village of Chakkchela and had been killed while riding his bike home from school. An investigation verified the many details Taranjit knew about Satnam's life. Satnam's family had kept several of his school notebooks, filled with writing assignments. Examination of Satnam's and Taranjit's handwriting (a trait experts know is as distinct as a fingerprint) revealed they were virtually identical.[47]

According to an article published in the *Indian Tribune*, author Jupinderjit Singh claims:

"If it is presumed that the soul is transferred from one person to another than it can be inferred that the mind will remain the same. Thus, if Satnam Singh's soul was believed to have been transferred into Taranjit Singh's body, then it stood reason that the handwriting of Taranjit Singh would correspond with that of Satnam Singh."[48]

Science is notorious for dragging its feet to consider new, novel, and "against-the-norm" concepts, which threaten its intractable paradigms. Ideas which threaten existing theories are considered aberrations, subject to rejection and ridicule. Those who think outside of the "scientific box" are frequently vilified as being mavericks, whose research is inconclusive and incompatible with traditional science.

Ignaz Semmelweis (1818–1865), Hungarian physician and

chief resident at Vienna General Hospital, is a compelling example. Vienna General had an excessively high maternal mortality rate (18%) due to puerperal fever, a bacterial uterine infection following childbirth. Puerperal fever was highly contagious, and quickly spread throughout the hospital. Extremely concerned, Semmelweis was determined to identify the cause of its transmission.

He observed that medical students would frequently autopsy cadavers prior to assisting in delivery rooms. At the time, the understanding of germ transmission was undetermined. However, Semmelweis theorized that students were transferring bacteria from cadavers to women giving birth. In other words, they were not sanitizing their hands after post-mortem examination. Thus, he insisted that medical staff wash their hands in chlorinated lime solutions after all procedures. Semmelweis's hypothesis proved accurate. Maternity ward mortality rates plummeted from eighteen percent to two percent.

Yet, the medical field ridiculed Semmelweis. Physicians refused to believe that their *own* uncleanliness could spread disease, and instead, upheld the prevailing theory that infection was spread through the air by miasma (a poisonous vapor or mist). Ostracized by his peers, he became increasingly depressed and isolated. His last years were spent in a mental asylum, forgotten by his colleagues.

Semmelweis' practice only earned widespread acceptance years after his death, when Louis Pasteur presented a theoretical explanation for Semmelweis' findings. Today, Semmelweis is known as "the man who invented modern infection control," and is widely considered *the* pioneer of antiseptic procedures. His discovery has helped to define modern medicine.

Yet, even today, traditional science remains highly skeptical of theories supporting the veracity of NDEs, OBEs, UAP-related contact, and PLR. Although skeptics assign various explanations for the above-mentioned phenomena, most believe they are confabulations, false memories, hypnotic suggestions, hallucinations, or products of overactive, creative minds.

Experiencers, including the authors, beg to differ. Almost unanimously, experiencers describe their encounters as "realer-than-real," life-altering, and utterly transformational. Research studies consistently maintain that experiencers believe consciousness exists beyond time and space, and that the brain is merely a receiver, rather than "container" of consciousness. The authors contend that extraordinary experiences are the consequence of the interconnected, entangled reality we exist in. As such, they share numerous commonalities. Let's examine these in detail.

# Chapter 5

## Convergence

What do NDEs, OBEs, UAP-related contact, and PLM experiences have in common? Could they be connected; tying us to the universe in multidimensional ways? Non-local consciousness may prove to be the link that explains it all. As Kenneth Ring asserts:

> When consciousness begins to function independently of the physical body, it becomes capable of awareness into another dimension. Let us, for ease of reference, simply call it, for now, the fourth dimension. Most of us, most of the time, function in the three-dimensional world of ordinary sensory reality ... When one quits the body, either at death or *voluntarily*, as some individuals have learned to do, one's consciousness is then free to explore the fourth-dimensional world. *Any* trigger that brings about this release may induce such experiences. It happens that coming close to death, for obvious reasons, is a reliable trigger affecting this release of consciousness. But to repeat: *Anything* that sets consciousness free from the body's sensory-based, three-dimensional reality is capable of bringing about an awareness of the fourth dimension. NDEers, OBEers, UAP-related contactees, and those with past-life memories seemingly transcend the time/ space/sensory constraints of our three-D realm, opening them to an awareness of the fourth dimension.[1]

Ring's analysis strongly implies that NDEs, OBEs, UAP-related Contact, and PLR are all interconnected phenomena and share many core components. Let's examine the most common themes reported by experiencers, which include but are not limited to:

1. Heightened senses
2. Information downloads (instantaneous transfer of data, guidance, or previously unknown information, most often, from what is referred to as "source")
3. Sensing a presence/otherworldly being
4. Telepathic communication
5. Time distortion
6. Otherworldly light

It is important to understand that these modalities do not always have a clear line of distinction. For example, many people describe NDEs as out-of-body experiences because in both cases individuals experience separation from the body. Many individuals also have UAP-related contact during OBEs and NDEs. Others experience past-life recall during OBEs and NDEs. Therefore, there are crossovers among the modalities themselves. Besides the six shared commonalities listed above, spontaneous healing and personal transformation will be discussed in following chapters, due to their significance and depth of information.

## NDEs
### Heightened Senses

Obviously, there is more to the human spirit than can be proved scientifically, and there is more to living than our sensory faculties define.[2]

P.M.H. Atwater

Universally, NDEers describe their experiences as "realer-than-real," or "impossibly real." According to Dr. Jeffrey Long, nearly 75% of NDEers indicate they have experienced "more consciousness and alertness than normal."[3] However, heightened lucidity or sensory input is inexplicable if consciousness is localized to the brain. How is *any* brain activity possible when clinical death is declared? Yet, NDEers consistently report extremely heightened sensory experiences after flatlining. As NDE researcher Pim van Lommel states: "We still do not know how it is possible for people to experience an enhanced consciousness ... during a period when the brain displays no measurable activity and all brain function, such as bodily and brain-stem reflexes and breathing have ceased."[4]

The two heightened senses most commonly discussed in NDE literature are enhanced sight and auditory perception.

### Enhanced Sight-NDEs

The most astounding aspect of heightened senses in NDEers is the omnidirectional (more than 360° panoramic vision) sight described by the congenitally blind. Van Lommel emphasizes, "The fact that somebody who has been blind from birth ... can nevertheless see [verifiable by third parties] people and surroundings raises significant questions ... This is impossible according to current medical knowledge."[5]

NDEer Vicki Noratuk, congenitally blind from birth, explains:

> I've never seen anything, no light, no shadows ... nothing. A lot of people ask me if I see black. No, I don't see black. I don't see anything at all. And in my dreams, I don't see any visual impressions. It's just taste, touch,

sound, and smell." Yet, during [my NDE], which was third-party verified, [I] "saw," with perfect, omnidirectional vision, for the first time. I was in Harbor-view Medical Center and looking down at everything that happened. The first thing I was really aware of is that I was up on the ceiling ... I looked down and saw this body, and at first, I wasn't sure that it was my own. But I recognized my hair ... it was long, down to my waist ... part of it had to be shaved off ... And then I saw my wedding ring on my left hand, and had my father's wedding ring on my right hand and a plain gold band that I had.[6]

How indeed, is it possible for the congenitally blind to possess an acute, 360° visual field, the ability to recognize and describe colors, objects, and the environment in fine-grained detail? These individuals apparently transcend all sensory restrictions. Does sight in blind NDEers depend on the eyes, or is there a non-retinol explanation? Researchers Kenneth Ring and Sharon Cooper have proposed a theory in which the spiritual body transcends limitations of the physical body and sensory organs. In their model, "The blind can see because, with the physical body temporarily inoperative, the spirit within them can use of the finer senses of the astral body, which presumably are perfect, to gain temporarily a kind of vision they could never have in life."[7]

### Enhanced Auditory Perception-NDEs

Research conducted by Dr. Joel Funk, professor of psychology at Plymouth State College in New Hampshire, revealed that approximately 50% of NDEers hear inexplicably beautiful, unearthly music. According to Dr. Funk, these tones were most often described as a New-Age synthesized genre. NDEers have described this music as transcendent, angelic; heavenly, joyous; sublime, coming from the core of one's soul. Others hear tinkling or Tibetan bells, or classical music, resembling the works of Beethoven, Strauss, or Wagner.[8]

Some NDEers report hearing the primordial sound of "OM," a combination of the Hindu words Atman and Brahman. Atman refers to the inner-self or soul, while Brahman signifies ultimate reality, supreme spirit, cosmic principles, and/or spiritual knowledge. It is believed that OM's vibrational frequency resonates with the entire universe. Research indicates that Tibetan bells, classical music, and OM have the same vibrational frequency of 432Hz (hertz). Hertz is a frequency measurement, based on vibrations per second. For example, one hertz equals 1,000 vibrations per minute. Unearthly music vibrates at extremely high frequencies (such as 432Hz) and is described as pure and undistorted.

> Music surrounded me. It came from all directions. Its harmonic beauty unlike earthly vocal or instrumental sounds was totally undistorted. It flowed unobtrusively like a glassy river, quietly worshipful, excitingly edifying, and totally comforting. It provided a reassuring type of comfort much like a protective blanket that whispered peace and love. I had never sensed anything like it. Perhaps "angelic" would describe it.[9]
>
> Dr. Richard Elby

Thus, otherworldly music envelopes NDEers with incredible love, peace, and joy. These harmonies are often described as "heaven's music," inaudible to our everyday, human auditory faculties. Experiencers explain these beauteous melodies as vibrations of divine, or universal love, expressed in a musical form.

## OBEs

OBEs offer a plethora of enhanced senses and emotions. Everything is grander, everything is more intense. When experiences occur beyond our physical bodies, closer to "source," we experience the trueness of the universe without our physical filters.

### Enhanced Vision and Colors

In our physical body, we can only see a small range of colors. I was astounded to enter into the astral and perceive a much greater range of colors than I ever had while in the physical. The perception of colors is based on light, and in the higher planes of the astral, objects seem to radiate (the air for example). On one occasion I left my body and requested: "Take me to a very beautiful place." I was whisked away to a place of such extraordinary beauty and color that I gasped over and over in disbelief and delight. It seemed like heaven has no place on earth matched its beauty or intensity. The colors were out of this world. Everything shimmered: the water, the air, everything! It was coursing with life and beauty. Returning to my body was a sour moment indeed![10]

Anne Moss, author, Astral Voyage blog

### Enhanced Auditory Perception

I was quite surprised to discover that OBErs can hear radio waves while in the magnetic body range! In my personal experiences, when I have left my body, I suddenly hear music and talking. If I turn a certain way or shift my focus, I pick up on the different frequencies in perfect clarity. I know this to be real for several reasons; one is that I hear classical pieces that are too marvelous for me to reconstruct in my own mind. I've heard exact words to songs that I only partially know. I believe I transcended time because I heard gorgeous songs that I know do not exist today. These are not classical or jazz pieces that I'm not familiar with, but songs one might hear on a soft rock station today. Gorgeous! If only I could write and sing, I would be famous! Recently, while coming out of sleep, (the hypnopompic state) I again heard a radio. This time I heard the time. It said, "Good morning, it's 8:00 o'clock" and at that moment, the downstairs bird clock chimed the black-capped chickadee which represented 8:00 o'clock. This so startled me that I forced myself awake and looked at my bedside clock to confirm it was indeed 8:00 o'clock! Why is this possible? I believe it's because our etheric body is also known as the electromagnetic body, and radio frequencies are affected by this.[11]

I've noticed a few times now when I'm about to experience an OBE, I've been able to hear things that I normally wouldn't have. For instance, I was taking a nap in my room upstairs while the radio was playing really quietly in my garage. I could decipher which station was playing (NPR) and could almost make out the words that they were saying, but when I "woke up" I couldn't hear the radio at all. I went downstairs to verify that the radio was on and I was correct. I understand that astral projection can make you hallucinate in a sense with visuals and sounds, but this felt different, and has happened on multiple occasions.[12]

Excerpt from r/AstralProjection

**Heightened Emotions**

OBEers report incredibly heightened emotions. As one OBE researcher explains:

When out of the body, you will love, fear, pity, more deeply than in the physical. I'm not sure what filter removes this sensitivity while in the physical, but emotions are enhanced in the astral. Perhaps when we build up "walls," they are, in part, due to our physical brain. It is said that our astral body is the formative dimension for emotions, and from what I've experienced, this is definitely true. I guess this is why they call the astral the "emotional" body.[13]

Lynn's personal OBE
May 19, 2011

I looked at the beauty of my surroundings. There were trees, shrubs, grass, and flowers. Every living thing was glowing with incredible colors. I lay down on my stomach to look at the grass and the dirt. I wanted to see more closely. The ground was moving in slight waves, as if it were alive and breathing. I lay there for a while ... It was breathtaking. My awareness was greater than anything I have experienced in the physical reality, but most of all I felt this sense of pure love and oneness with my surroundings. I was no longer me, but everything around me. I rose and began flying into the sky.

I asked for greater awareness. I said, "Please take me where I need to go." I began to move but was suddenly blocked by a barricade of some kind. It was huge, round, and black. Everything black, yet, I could see the stars. I heard loud beautiful music playing loudly in all directions. I didn't recognize it, but it was exquisite. Something told me to pay attention to the barricade. As I stared at it, symbols began to appear, etched into the barricade itself. I could not understand them. They may have been written in an ancient language. They were both beautiful and elegant and took my breath away.

## NHIB/UAP Contact

*Image 5-1*
*Illustration of Unidentified Aerial Phenomena Abduction*

### Bodily Sensations/Vibrations/Sleep Paralysis

Abductees (while in bed or asleep) commonly experience strong vibrations or electrical shock sensations moments before extraterrestrial contact. The majority of researchers maintain these sensations are related to sleep paralysis, rather than UAP-related phenomena. Meta-physical blogger and UAP-contactee Farusha, however, begs to differ.

As he explains: "I have had various episodes of sleep paralysis since childhood that were both exceedingly terrifying and interesting, however, most memorable episodes had an additional element—that of a very fierce and painful vibration in the head and the face, particularly (in my case) in the area around my nose and hypothetical third eye (between the eyes) ... Then my face and head would begin to buzz dramatically."[14]

From his numerous personal experiences, he postulates:

> A human being (and most mammals) are
> at their most vulnerable while asleep. They are
> even more vulnerable when they are in a state
> of "sleep paralysis." What if entities, good,
> harmful or neutral, were aware of the human
> mammalian function of sleep paralysis and
> could cause it at will in any susceptible human?
> That would explain many abduction accounts
> and also accounts of spiritual revelation. It is
> possible that all persons who experience the
> paralysis and the vibrations are not receiving
> contact from the same beings. According to
> mathematicians at least eleven dimensions
> exist in our Universe, though we are sensorily
> aware of only four. In addition, whole parallel
> universes have been postulated by some
> scientists. The vibrations would seem to be a
> crucial part of the ability to receive
> communication from these beings who may
> inhabit another dimensional plane. They may
> need to change our vibration to attempt to
> communicate with us or transport us into
> another time-space continuum.[15]

Lynn believes that most contact occurs on the ethereal
plane. She explains her understanding of the interconnection
of vibrations and communication:

> This is where we find ourselves during
> OBEs, after the vibrations and sleep paralysis
> have subsided. Others speculate that beings
> can bring us into the ethereal plane. If this is
> so, then the prelude to this would be to put us

in the vibrational state where sleep paralysis usually occurs. This may happen mostly during the night. Although, some may find themselves slipping into a meditative state from wakefulness, having an inter-dimensional shift. However, I feel this is a two-way communication and that the entities are on some type of a familiar basis with us. I also believe that we can initiate contact with other dimensional beings.

**Sounds**

UAP-related contact does not include "enhanced hearing" in the traditional sense. Rather, individuals describe unusual noises which do not occur in their everyday lives.

Numerous people who report UAP sightings hear an accompanying humming, whirring, or buzzing noise. Some have termed this "The sound of a UFO." For nearly a decade (between the late 1980s to 1990s), hundreds of residents in Taos, New Mexico, reported hearing unusual humming sounds. In fact, this noise became so prevalent and frequently noted, it was nicknamed the *"Taos Hum."* During this period, coincidently, UAP sightings proliferated in the region. An investigation was led by U.S. Representative Bill Richardson to determine if the *"Taos Hum"* was related to secret experiments conducted by the Department of Defense. The investigation failed to uncover a satisfactory answer. Many Taos residents insist the Hum is directly related to UAP sightings.

Numerous researchers, however, maintain the Hum is a spontaneous otoacoustic emission (SOAE)—a sound generated from within the inner ear. SOAEs are not considered an auditory disorder, such as tinnitus. Studies report that between 38%–68% of adults with normal hearing occasionally experience faint SOAEs in quiet, or completely silent environments. Many report SOAEs as short-lived ringing, buzzing, or high-pitched "screeching in the ear."[16] However, humming and buzzing noises associated with UAP have been reported worldwide. A sighting with accompanying humming noises, occurring in Ponce Inlet, Florida, was reported to the Mutual UFO Network (MUFON). According to MUFON:

> At 10:15 pm on September 14, 2015, numerous witnesses in Ponce Inlet, Florida, reported loud humming sounds. The sounds were accompanied by a hovering, low flying object surrounded by lights. (Case 59867 from the Mutual UFO Network (MUFON) witness reporting database). The reporting witness heard the humming sound for five to eight seconds and first thought it might be someone's garage door opening. "But it was unusual, unlike anything we ever heard before," the witness stated. "It was so loud it caused alarm. So we immediately went outside. Several neighbors went outside to express concern and alarm. The dogs in the neighborhood were barking and then they got very quiet."[17]

According to the CIA, the case of patrolman Lonnie Zamora (1964) has been called "The best-documented case on record" ever investigated by the U.S. Air Force's famous UFO program, Project BLUE BOOK. Zamora was "chasing a speeding vehicle when he heard a loud roar and saw flames near the area of a dynamite shack. He abandoned the chase and drove through considerably tough terrain to investigate ... He came upon a UAP roughly the size of a car. Stopping his patrol car, he approached the object on foot and began to hear highly unusual sounds. He later described the noise as 'The

100

noise went from a low pitch to a high pitch. It was unlike anything I'd ever heard."[18]

From 1958–1968, the National Investigations Committee on Aerial Phenomena (NICAP) examined one hundred three cases of UAP-related contact noise. Interviewees reported five categories of unusual sounds. These included violent noises, humming, a rush of air, high pitched tones, and coded signals. Violent noises, described as an instantaneous explosion like a shock wave, or loud, continuous sounds were specifically related to cases in which a craft landed. Many case studies reported a buzzing noise similar to a swarm of bees. High pitched sounds were described as "shrieking," "a piercing whistle," or a "whine." Numerous interviewees related high pitched noises to that of a jet aircraft as the engines are revved up. Others described hearing a "loud turbine" or "high pitched scream of a jet engine." Coded signals were most frequently referred to as "beeping noises." One case describes the sound like a "modulated whistling."[19]

Thus, the question remains: Is this merely a commonplace auditory occurrence, an inexplicable mystery, or the actual "sound" of a UAP?

### Past-Life Recall
### Heightened Emotions/Sensory Perception

Ineffable. Intense. Powerful. The heightened sensory perception experienced during past-life recall is nearly inexpressible. I can best describe it as a multi-dimensional, emotionally impactful holographic smorgasbord of the senses.

Barbara

Past-life recall has a "cinematic" feel. Imagine you are an audience member watching an intensely emotional movie. The film is a "biography" of *your* life, occurring in another time and place—the intense portrayal of a life once lived, featuring the emotional scars, the fears, and complexities of that *particular* life. The movie ends. You leave the theater deeply affected by both the film and performances—touched to the very core of your being. Carol Bowman, therapist and past-life researcher, describes past-life regression as:

> An amazing, full-sensory experience. You might experience the memory as a vivid movie or see only vague flashes of images that prompt the narrative. You might hear gunshots or explosions on a battlefield or music at a dance. It is possible to recall smells too: smoke from a fire, leather from a saddle, or the sweat of a dirty body. As the story unfolds, you feel real emotions appropriate to the story. You may cry when you re-experience deep sadness at the death of a beloved child, feel despair in the pit of your stomach as you witness a massacre, or elation at a long-awaited homecoming from war. And just as you can recall strong emotions, you feel the pain of an arrow piercing your body as you are dying, or the heaviness of a load you're carrying on your back. These physical sensations and emotions are very real at the moment, but pass quickly as you move through the past life story and death.[20]

Dr. Brian Weiss is a world-renowned past-life hypnotherapist and researcher. He describes past-life recall in a similar vein:

> In hypnosis, the person is the observer as well as the person being observed. In fact, many people in the trance state watch the past

as if they are observing a movie. Your conscious mind is always aware of and observing what you are experiencing while you are hypnotized. The hypnotized mind, *always retaining an awareness and knowledge of the present,* puts the childhood or past-life memories into context. For example, if the year 1900 flashes, and you find yourself in ancient garb building a pyramid in ancient Egypt, you *know* that the year is B.C., even if you don't see those actual letters.[21]

How is one able to access past lives under regression, or in rare cases, spontaneously? How is the brain, a three-pound organ, or a mere two percent of our total body weight, able to transcend the impossible vastness of time/space? Perhaps the brain, under certain conditions, is able to vibrate or tune into the same "frequency" as universal consciousness. This vibrational state may allow the retrieval of lives from a non-physical plane of existence (or etheric plane), also known as the Akashic Records. The brain produces five primary types of brain waves: beta, alpha, theta, delta, and gamma. Numerous PLR clinicians propose that past-life regression occurs when a client enters the theta brain state.

As Clinical Hypnotherapist William Hewitt explains: "The Theta range of brain activity is approximately four to seven cycles per second. It is during the Theta Brain Wave State that a portal or pathway between the conscious and unconsciousness mind occurs, and where deep transformation takes place at a subconscious level. Additionally, it is considered to be the psychic range of the mind—the area where psychic experiences occur."[22]

Thus, past-life recall evokes powerful and visceral emotion. Although the intense *feeling* is an inherent feature in the recall, enhanced sight is frequently described as the most *extraordinary* sensory feature. As in NDEs and OBEs, vision in past-life recall is not limited to three dimensions. Experiencers describe seeing a world in detailed "multi-vision," (the ability to see from all angles instantaneously: up, down, sideways, above, below, and behind). Multidimensional sight transcends the limitations of human vision. Researchers Kenneth Ring and Sharon Cooper refer to this as more than 360° vision. Everything, every single detail, is seen with crystal clarity.

### Information downloads

#### NDEs

Dr. Jeffrey Long, leading NDE researcher, conducted a study with over thirteen hundred participants. His statistical analysis concluded that 56% percent of NDEers reported "Knowing special knowledge, universal order, or purpose." Approximately 32% stated, "I seemed to suddenly understand everything." Over 31% noted, "I seemed to understand everything about myself or others."[23]

Near-death experiencers overwhelmingly speak of knowledge gained as spiritual, universal, or pre-cognitive in nature. Hafur, one of Jeffrey Long's NDERF survey respondents, describes information gained from her NDE:

> Death is a metamorphosis of time—one more illusion born of our mental concepts. Consciously living by love is the essence of life itself. We live in a "plural unity" or "oneness." In other words, our reality is "unity in plurality and plurality in unity."[24]

Kimberly Clark-Sharp (Co-founder and President of the Seattle International Association for Near-Death Studies) describes the knowledge she received during her NDE:

The Light gave me knowledge, though I heard no words. We did not communicate in English or in any other language. This discourse was clearer and easier than the clumsy medium of language. It was something like understanding math or music—nonverbal knowledge, but knowledge no less profound. I was learning the answers to the eternal questions of life—questions so old we laugh them off as clichés. Why are we here? "To learn." "What's the purpose of our life?" "To love." I felt as if I was re-remembering things I had once known but somehow forgotten, and it seemed incredible that I had not figured out these things before now.[25]

Additionally, NDEers may be informed of future events. However, some individuals may receive profound scientific or mathematical downloads. Olaf Swenson and Lynnclaire Dennis are two such examples.

At the age of fourteen, Olaf Swenson had a NDE during a surgical procedure. As he explains: "I had total comprehension of everything. I stood at the annihilation point, a bright orange light. As I felt my mind transported back to my body, I thought, please let me remember this new theory of relativity." Later in life, he created over one hundred patents in molecular chemistry.[26]

Lynnclaire Dennis lost consciousness during a high-altitude hot-air balloon ride. During her NDE, Lynnclaire received one of the most complex mathematical downloads ever recorded, when she was shown the structure of God. Dennis calls God "The Pattern," and describes the Pattern as: An all-connecting pattern of Light—a three-dimensional mandala, representative of time, space, and the energy generating matter. The Pattern is a knot; but not just any knot. The Pattern is the most profound knot—a never-before-discovered trefoil knot. The Pattern became of great interest to mathematicians, physicists, astrobiologists, and molecular geneticists who developed from it the "Mereon Matrix"—an algorithm representing the unification of knowledge which relies on whole systems, both living and life-like. It is a scientific framework charting the sequential, emergent growth process of systems.[27]

*Image 5-2*
*Flower of Life Sacred Geometry*

This type of pattern is known as sacred geometry. Geometric computations of precise measurements are seen in nature. The golden ratio, also known as the Fibonacci sequence, is found everywhere in the universe. It is seen in the chambered nautilus, flowers, plant growth, animals, environmental structures, both ancient and modern architecture, and the rotation of planets and galaxies. This is a true mathematical formula that that is found in relation to all living things, and the creation of life itself.

As Lynn explains:

> When in a type of hypnogogic state, I have often seen beautiful, colorful, rotating geometric patterns. Because I see them in this state of mind, I often wonder if they play a role in inter-dimensional consciousness. Could these be the holographic, mathematical

106

formulas that we see as physical reality? I have spoken to others who have also seen these patterns during altered states of consciousness. Mind-altering drugs like LSD and magic mushrooms are known to produce these geometric patterns. Perhaps these drugs induce a state of consciousness in which one is awake, yet sees these patterns, or elements, that connect us to other dimensions.

*Image 5-3*
*Fibonacci Geometry Mathematics*

**OBEs**

OBEers may experience various types of information downloads. Experiencers commonly talk about a type of training or education taught by guides and teachers during astral projection. We coin this term "night school" and it normally occurs at a type of university in alternate dimensions. I myself have experienced this on several occasions. It seems that information occurs through experiences, rather than receiving an instant downloading of information. Not all of these experiences are remembered during wakefulness, but they come to the surface as memories at various times.

Lynn

Robert Monroe spoke of these classes in his book *Far Journeys*:

With the mutual recognition of such communication, the depth and extent of my OOB (out of body) patterns shifted. I was escorted frequently to what might be loosely described as another kind of class, in that there was an instructor and there were students, including me. It was entirely different from the sleeper's classes I remembered. Here, freely translated, there was a brilliant white, radiating ball of light that was the teacher. I could detect radiation of others—presumed students—all around me, but nothing beyond that, no form or any indicators as to who and what the others were. Instruction consisted of a seeming sequential bombardment of packages of total experiential information to be absorbed instantly and stored thought balls, whose actual name cannot be translated into a word, which I called rotes. It apparently is a very common communication technique in NVC (non-verbal communication). What I could bring back, I attempted to convert into human, not inhuman usage, with mixed results. I have been unable to relate the vast majority of such information in any way to life here on time-space earth. It may be a preparation for activity yet to take place here, for use in other non-physical energy systems, or it is beyond my comprehension. The last is most likely.[28]

Like Monroe, Lynn has received advanced telepathic information during her OBEs. As she describes:

> I can relate to Robert Monroe. The information received during an OBE is often beyond human comprehension. For example, during OBEs I have perceived great gates and walls full of inexplicable symbols. I have been shown advanced mathematical equations. At the time, I understand their meaning, although, in full consciousness, they escape me. I have had glimpses of being someone else and doing extraordinary things in other dimensions. During many OBEs, I am being trained by my guides and teachers. The knowledge I gain lays just beyond my conscious state. Yet, this insight has changed the very core of who I am. Perhaps I will never completely understand this "education" until I have left this world and its mental constraints.

### UAP-related contact

Contact experiencers tend to receive downloads regarding science, physics, technology, spirituality, philosophy /metaphysics, human behavior, and ecology. Extra-terrestrials and Inter-dimensional beings not only teach and communicate through symbols, they also use more direct approaches such as images.[29] As above-mentioned by Robert Monroe, numerous contactees receive this downloaded knowledge via "night school."

Whitley Strieber, novelist, and contactee describes being abducted in San Antonio, Texas, in the summer of 1954. Strieber's experiences at a secret nighttime 'summer school' serve as the basis for his book, *The Secret School* (1996). Nine-year-old Whitley was whisked away to a secret nighttime 'summer school' located in San Antonio's wild Olmos Basin. Here, he and a group of other children received a series of nine lessons, focusing on time-travel and connection with the universe. The lessons were conducted by a nun-like being, named Sister of Mercy. The children were given a type of virtual reality helmets, which, according to Strieber, enabled them to view future events. Strieber relates his ninth lesson: "I saw a flat-screen TV, obviously unlike any kind of TV available in 1954. The TV was switched to a news channel, and on the screen, I saw a number of scenes that have remained in my mind all of my life—not exactly as a conscious memory, but rather as a reservoir of visual images that I have come to draw on in my work." Some of the scenes were of catastrophic events that appear to have since come true, such as the Great Malibu Fire of 1993."[30]

Mary Rodwell, renowned hypnotherapist, author, researcher, and founder of the Australian Close Encounter Resource Network (ACERN) asserts:

> Many experiencers mention space schools, the encouragement of a multi-awareness and advanced maturity, knowledge bombs and a deep emotional connection to the 'galactic family' concept. Many of them are highly frustrated because they are not, as yet, consciously aware of what their 'mission' or

purpose is. There is, however, an awareness of being taught something, of being shown things through the mind, and there is an increased consciousness regarding energy, frequencies, and how to manage them.[31]

Experiencers tend to have a deep longing for, and connection to their galactic family (a group of physical and non-physical beings who are inter-related energetically and/or physically to human beings). Many consider themselves strangers on earth and express a strong desire to return "home." Numerous contactees feel a great longing for loved ones not presently living in our earthly dimension. They often mention having vague memories of living other, extraordinary non-earthly lives, yet, these memories lay beneath daily consciousness. Experiencers believe that part of our earthly journey is to overcome these boundaries that separate us from the higher dimensions.

**In Past-Life Recall**

In past-life recall, previously unknown data/information is not accessed via what is commonly referred to as a download. Einstein's theory of relativity is not instantaneously bestowed upon us, nor do we typically receive complex scientific, mathematical, or technical data. Rather, past-life recall is a *remembrance*—a *recollection* of *how* and *why* a particular life was lived, its very purpose. Recalling a past life gives us the opportunity to resolve/ understand karmic debt, unfinished lessons, and the nature of duality.

Our task is to learn, to become God-like through knowledge. We know so little ... I have so much to learn. By knowledge, we approach God, and then we can rest. Then we come back to teach and help others ... A life cannot be worked on a schedule as so many people want it to be. We must accept what comes to us at a given time, and not ask for more. But life is endless, so we never die; we were really never born. We just pass through different phases. There is no end. Humans have many dimensions. But time here is not as we see time, but rather lessons that are learned.[32]

Catherine
(last name withheld for anonymity)

In my most current past life, I had no regard for my physical health. On a daily basis, I smoked like a chimney and consumed alcohol as if it were water. In my present incarnation, I eat well, exercise, abstain from cigarettes, and rarely drink alcohol. Regardless, I have struggled with a lifetime of challenging health issues (thanks a lot, past life!). Yet, I understand why. Many would call this karmic debt. I like to refer to it as "an experience in duality." I did not learn self-care in my previous lifetime—I have been given the choice to acquire this knowledge now. Our physical bodies are the vehicles with which we navigate this three-dimensional existence, and they must be loved and cared for. They are the vessels of our eternal consciousness and it is through them we experience and grow.

Barbara

Recollection gives us the opportunity for spiritual growth; the chance to "start over where we left off last time." The ultimate purpose of past-life recall is knowledge—learning to become wiser in our current incarnation. Yet, how do we retrieve these memories from another time, place, or dimension?

As psychiatrist and past-life researcher Jim Tucker explains, the key to unlocking the mystery of past-life recall may lay in the foundation of quantum physics: "The answers might be found within the foundations of quantum physics, which demonstrates that the physical world is affected by, and even derived from the non-physical, from consciousness. If that's true, then consciousness doesn't require a three-pound brain to exist, Tucker says, and so there's no reason to think that consciousness would end with it. It's conceivable that in some way consciousness could be expressed in a new life."[33]

Dolores Cannon, the late past-life regressionist, developed a hypnotic technique she termed the "Quantum Healing Hypnosis Technique," or QHHT. Much like Tucker, Cannon believed past-life memories are retrievable via quantum mechanics. Her novel technique has been described as:

> A powerful tool to access that all-knowing part of ourselves that has been called The Higher Self, The Oversoul, even the Soul itself. When we incarnate on Earth, we forget our previous lives and connection to our souls and The Source. QHHT enables all people from any background, culture, religion, or belief system to engage with what she calls The Subconscious since it resides beyond the conscious mind. Dolores's term The Subconscious, which she later abbreviated to The

SC, is that greater part of ourselves that is always connected to The Source, or God, and has unlimited knowledge and an unlimited ability to heal the physical body. Sometimes mental and physical ailments are rooted in trauma from past lives; sometimes they are connected to lessons being learned in a person's present life. The SC reveals the cause and will assist according to any soul's particular lessons.[34]

### Sensing a Presence or Being(s)

#### NDEs

In a study conducted by NDE Researcher Dr. Jeffrey Long, 57.3% of survey respondents encountered mystical beings or deceased friends/relatives. In a study published by Pim van Lommel in the esteemed journal *The Lancet,* 32% of cardiac patients met deceased friends or relatives. Renowned NDE researchers Bruce Greyson and Ian Stevenson studied seventy-eight NDEers. Their findings concluded 48% of subjects reported meeting some person(s) not physically present, including a being of light; 25% encountered religious figures; 16% met deceased acquaintances; 14% saw living acquaintances; and 26% met unidentified strangers.[35]

Thus, beings encountered by NDEers fall into two main categories: either a luminous being of light, or those who present in human form. Light beings may appear in different forms according to an individual's religious or cultural background. Christians may refer to them as Jesus or Christ, while others term light beings as God, Angels, or Source. Regardless of presentation, light beings are described as having a distinct personality and exuding indescribable love and acceptance. Universally, beings emphasize the importance of loving others and continuing to acquire knowledge throughout one's lifetime.

A Christian NDEer described meeting a Christ-like being: "I could see this light. It was a very brilliant light ... I was trying to get to that light at the end because I felt that it was Christ ... For immediately, being a Christian, I had connected the light with Christ, who said, "I am the light of the world."[36]

One experiencer encountered human-type beings: "I was surrounded by other beings, or people, who I felt as though I recognized. These beings were like family, old friends, who'd been with me for an eternity. I can best describe them as my spiritual or soul family. Meeting these beings was like reuniting with the most important people in one's life, after a long separation. There was an explosion of love and joy on seeing each other again between us all."[37]

NDEer Ned Dougherty met spiritual beings, relatives, and deceased friends. As he describes: "I turned to my right, realizing that a group of spiritual beings had joined us on the celestial field. This event was indeed a homecoming for me. Among the group of spiritual beings, I recognized deceased friends and relatives from my life. I also recognized other friends from my spiritual life prior to my birth on earth. I was filled with joy when I recognized my grandparents, aunts, and uncles who had died during my life. However, I was disappointed because I did not see my dad among the group. I then recognized other relatives and friends from my life, including a girl from high school.

I did not know she had died."[38]

Barbara's father had a similar experience as he was nearing death:

After a long, arduous battle with pancreatic cancer, my father was placed in hospice care at home. It is important to note that he was not medicated and completely lucid when the following events occurred. As he neared death, he had several encounters with light beings. Nearing-death experiences with beings frequently mirror those of actual NDEs. One morning, dad urgently called me to his bed. He insisted I grab a paper and pencil and write down EXACTLY what he said, "Earlier this morning I saw a being, not a person. It definitely was not a human being. Together we hovered above the foot of my bed—we were looking down at myself on the bed. This glowing being told me a party was going to be given in my honor, and I was to be the special guest." Days later, he described several beings of light, smiling with joy and kindness, as they prepared him for the Jewish (dad was a non-practicing Jew) cleansing ritual. This ritual cleansing bath, termed *Mikvah*, is the purification rite of a male Jew before he is laid to rest. These light beings filled dad with peace and joy.

Light beings are universally described as gentle, infinitely wise, possessing definite personalities, and unconditionally loving. Virtually all NDEers state that the degree of love and joy felt in their presence is inexplicable in human language. Largely, NDEers believe light beings serve as spiritual guides, whose purpose is to ease their transition into the afterlife, or non-physical realm. NDEers commonly refer to this transition as a joyful "homecoming" or a return to their true existence.

### OBEs/UAP-Related Contact

Numerous individuals report seeing non-physical beings during OBEs. These beings often appear as "spirit guides," watchers, helpers, or other astral beings. Many of them are human or humanoid in appearance. Others believe that they are being visited by UAP entities, often reporting seeing them appearing next to their beds.

William Buhlman's survey revealed that 22% of the respondents reported having seen or felt the presence of unknown non-physical beings, and 24% stated that they have seen, heard, or spoken to a deceased loved one.[39]

The late psychiatrist John Mack contended: "The alien beings function as spirit energies or guides, serving the evolution of consciousness and identity."[40]

> How we interpret this presence is of course based on our individual beliefs and perceptions of reality. I have heard descriptions that span the limits of imagination: angels, devils, aliens of all sorts—even animals. The most commonly reported sightings are humanoid forms that appear to be watching as the out-of-body experience unfolds. We frame our nonphysical adventures and contact based on our individual mindset.[41]

Not all individuals who have OBEs observe NHIs. Many report seeing deceased loved ones, friends, family, animals and pets. They may also see friends and family who are currently living. Family members, no matter their age or physical condition at the time of death, appear perfectly healthy and in their prime (usually in their 30s).

March 3, 2013

Experience from Lynn Miller's OBE journal

I was on the verge of having an OBE, focusing on increasing my vibrational state. I concentrated intently on my white noise machine, taking this external rhythm and matching my internal vibrations to it, with the intention of increasing my vibratory state. I felt my heart beating rapidly, mimicking the sensation.

This began as a false awakening, because although I remembered the vibrations, I did not remember exiting out of my bed the way I normally do. I performed a reality test several times, noting that my hand went through the wall ... I scratched at the inside of the wall and felt the drywall chalky stuff in my fingernails.

I exited the bedroom and I ended up at a different house, not the condo where I was living in at the time. Mom was there, yet it was not her house either. I was with a guide of some kind, a younger Hispanic man. (My guides seem to be Hispanic; I wonder why?) He walked me to the outside of the house, out of the front door. As I walked out the door, my dad came to me. It was as if my guide was taking me to him or arranging this meeting. I was immensely surprised. "Daddy!" I yelled; all we did was just hug each other in this tight embrace. He was much younger; how he looked I would say in his 40s, long before he had his strokes and lost weight. We both just stood there, laughing with pure excitement. I think I was crying and laughing at the same time. The embrace lasted several minutes. Then I asked him questions, yet do not remember what I asked. I wanted him to see mom, but my guide said he had to go back. The guide was explaining to me why my dad had to go, but I do not remember what he said. After he left with my dad, I went back into the house. During this time, I felt myself slipping away; I was walking through the house, then I stopped and demanded more clarity and awareness. I looked at my hands to concentrate—this time I had stubby fingers, and I counted six! I kept walking through the house trying to concentrate on the surroundings. I was looking for mom; I thought I saw her, but I am not sure.

I exited the house and went outside. I looked up at the sky and I started to talk to the ETs, asking them, please show me what you think I need to see. I jumped up into the sky, but I felt heavy, and I started to fly backward. I felt heavy though and I remember swooping down and feeling the sensations in my stomach. I thought I was heading to something specific, but I woke up instead. I think that I already saw what they intended me to see ... it was my dad.

The experience of seeing my dad was unexpected, yet astonishing. In the past I had made several attempts to see my dad in the afterlife during an OBE. I once saw him in the past as a young child. Experiencing this was one of the most significant occurrences in my life—knowing that dad had not only evolved to his higher self but was also there watching and helping me.

## Past Lives

Beings are rarely encountered during spontaneous past-life recall. They are most frequently met during a Life-Between Lives regression (LBL). While in a super conscious state, some individuals regress to the period immediately before their current incarnation. This is referred to as LBL—an often life-altering experience occurring in the spirit realm, between incarnations. In this state, beings most often present as spirit guides, soul groups, or the council of elders. LBL regressionist Susan Wisehart explains the roles of the council of elders and soul groups. As she states: "The meeting with the council of elders is a compassionate and unconditionally loving review of your most immediate past life and your current life to better understand soul lessons, life purpose, and healing of repeating karmic patterns. The meeting with your soul group assists in understanding the role that different souls play in your life, such as soul mates."[42]

An LBL regressionist (who prefers to remain anonymous) details one of her sessions:

> Before I [client] know it I am being guided
> to a place of healing. Here I am led into an area
> where concentric circles of light are directed
> over and through me. Much like sound waves,
> these circles of light heal me of the "density" I
> have brought with me from my life on earth. I
> stay here for some time and don't really want
> to move on. In time I feel rejuvenated and
> know it is time to go. The next stop in my
> journey is before the Elders or Council of Light
> Beings. As I approach I am feeling guilty and

apologetic about not having lived my life well enough. The message comes loud and clear "There is no place for guilt here" and "You have done well."

I am greeted by five light beings wearing energetic robes of light. They emanate the most beautiful bluish white light and I am overwhelmed with total and unconditional love. Tears flow now as my life is reviewed and my progress as a soul is discussed. All of this is done in a most loving and accepting way. My life patterns and soul lessons are shown to me and I am taught about judgment, separation, and the interconnection of all beings. I am also able to ask questions about my life's purpose and some of my experiences and lessons. This will all help me live my current life in a much more conscious and meaningful way.

From the council chambers, I am taken to a wonderful gathering place where I can sense the presence of those souls with whom I have deep and lasting connections. They seem to line up in rows waiting to greet me. There is an indescribable sense of being welcomed back from a long, long journey away. It is a joyous reunion of souls.[43]

Telepathy

**Telepathy, from the Greek tele ("distant") and patheia ("feeling")**

Telepathy refers to the ability to mentally transmit (telepathic communication) or receive (telepathic perception) information non-verbally. Telepathic communication is the ability of humans to communicate information from one consciousness to another, without the use of speech or body language. "Telepathic abilities are about connecting frequency. It is like turning on a radio and finding the right station. You just have to know how to 'tune in' and the frequency of the program."[44]

### Telepathic Communication

Psychic medium Michael Robey asserts that:

Telepathic communication is the creation of thought, which is of a high energetic vibration: a higher oscillation than sound or light vibration, permeating through the Universal Energy of Frequency and Vibration Field. By intention, you can focus the telepathic communication to one or several specific conscious being(s), whether these are human spirits, animal spirits or other forms of spirit beings, existing (i.e. vibrating) on the third dimension (e.g. alive on this planet), or existing (vibrating) on a higher level of vibration (i.e. nonphysical). This communication becomes an energy, which is imprinted into the ether, where there is no time or space. In turn the communication can be instantaneously received by the receptor, regardless of his/hers/its whereabouts or existence on the physical third dimensional level, or indeed a higher dimensional level. Just as you can telepathically communicate, you can also receive telepathically.

Telepathy is the universal language used to communicate during NDEs, OBEs, UAP-related contact, and past-life recall. Information may be relayed in various forms, including mental imagery, or via emotion.[45]

## NDEs

NDEers consistently describe conversation in the non-physical, afterworld realm as telepathic. Researcher Vince Migliore conducted a research study with over seven hundred NDEers. His statistical analysis (which includes telepathy) was published in *A Measure of Heaven: Near-Death Experience Data Analysis* (2009). Migliore concluded that: "People hear voices during their NDEs, but this communication is a direct, mind-mind communication, and a mechanical vibration of sound. Those who report telepathic messages often note how precise the method is compared to the use of words. There is no confusion or doubt about the concept being conveyed."[46]

I would 'ask' then would 'know' the answer from the golden, glowing, loving being. I had no lips to speak and no ears to hear, but I heard and spoke somehow. So did it. I reveled in that complete, pure, communication. There was no possibility of misunderstandings or evasions. There were no words to confuse the issue, only the truth of learning and knowing each other between us.[47]

Information was relayed from my grandmother's spirit directly to mine. It imprinted itself on me without any effort on my part to understand or remember it—it became part of me.

Penny

Dr. Raymond Moody, commonly known as the "Father of NDE Research" further states:

This unimpeded exchange does not necessarily take place in the native language of the person (NDEer). Yet, he understands perfectly and is instantaneously aware. He cannot even translate the thoughts and exchanges which took place while he was near-death into the human language which he must speak now, after his resuscitation.[48]

Perhaps the most remarkable telepathic communication is reported by profoundly deaf individuals.

Suddenly, I felt myself being pulled very swiftly through total blackness to the entrance of a churning tunnel, its walls reflecting the neon-blue color that my non-corporeal body was emitting. I still had all of my faculties and feelings ... A low, steady droning sound filled the narrow passageway, this sound coming not from the walls but from me. As a profoundly deaf individual, hearing this sound was beautiful, since it brought comfort as would a mother's voice for a child.[49]

### OBEs

My guides have always been with me, not just in this life, but over multitudes of lifetimes. They are my family, my friends, my teachers. Life on planet earth is difficult. We are here, and here we have this veil of forgetfulness and the feeling that we are separate from all things. We are in a denser state of reality, a lower frequency of being. My guides making their presence known to me seemed more like a metamorphosis. It was slow at first. When I first saw their presence, it was like looking at a silhouette of their form. A type of energy form. As my awareness and thought frequency increased, I was able to see them. I was confused at first because no words were spoken. I am an earthling and here we use our mouths to speak. It was a huge paradigm shift for me when I first began to understand how communication manifested inter-dimensionally. You can call it telepathy, but the best way to describe it is that it is a "knowing." Our thoughts become entangled and merged into one. When I first experienced this, I went into fit of frenzy. I said, through spoken words (well at least it was my spoken words through my astral body), "You know me! You know all about me! You know all my darkest secrets and all my fears, and every terrible thing I have done ... but you know me, and love me, and you do not judge me!" After that, I felt complete transparency with them. During my journeys to other worlds, to other dimensions, to other races and civilizations, it has always been with complete transparency. There is no deceit, there are no lies, there is only truth.

Lynn

### UAP-related Contact

Telepathy is the most constant feature reported in UAP-related contact. Numerous experiencers believe the unique physiology of non-human intelligent beings inhibits their ability to verbally communicate. For example, insectoids (insect-looking beings), are often described as having "slits" for ears with negligible mouths and noses. Additionally, gray beings, whether tall or short (ranging from approximately 3 ½ - 6 feet tall), are frequently reported as having no ears, a "slit-like" mouth, often with no teeth, tongue, larynx, or respiratory system; thus, lacking the basic aspiration required for sound formation. The significant physiological differences between human and alien may *marginally* explain the necessity for reported non-verbal or telepathic communication.

However, it is far more likely NHIs "speak" from higher levels of consciousness. In advanced, multi-dimensional realities, there aren't any barriers between thought and understanding, or communicator and receiver. The idea that communication is restricted to verbal interaction is a purely human construct. Communication in the "3D" world of concrete understanding is restricted to our five senses. Telepathic communication, however, is communicated via the sixth sense, in an instantaneous exchange of ideas, knowledge, and thoughts. It is a manner of conversation in which misunderstanding/ misinterpretation is nonexistent. To consider the very concept requires "thinking outside of the box."

Let's suppose advanced beings vibrate at extremely high frequencies, while we "less evolved" humans vibrate at considerably lower and denser frequencies. A normal human can hear between 20 Hz and 20,000 Hz. (20 Hz is an extremely low pitch, or deep tone, while 20,000 Hz is an exceptionally high tone).[50] Most of us are familiar with the proverbial dog whistle—the whistle that dogs (and yes, even cats) can hear, which is inaudible to the human ear. Dogs can hear frequencies higher and outside of the normal range of human hearing, between 47,000-65,000 Hz. Elephants, however, can hear frequencies that are too low for the human ear to detect. The lowest calls, the rumbles, often have fundamental frequencies in the infrasonic range (fundamental range from 5 to around 30 Hz) and are able to hear sounds from up to 5 miles away.[51] If we creatively "guesstimate" that NHIs vibrate at frequencies between 0 Hz and 100 Hz, sensory communication between human and NHI would be impossible. Is it any wonder then, that telepathy is the *universal*, inter-dimensional mode of communication between *all* contact modalities?

## Past-Life Recall

Telepathic communication in past-life regression differs from that in NDEs, OBEs, and UAP-related contact. Experiencers refer to this phenomenon as "body wisdom," or the ability to comprehend non-verbal communication via sensing and feeling. As with telepathic exchanges, sensory information and images are transferred to experiencers directly. During past-life regression, individuals enter a deeply relaxed state in which they are able to release the analytical tendencies of the left brain. Research demonstrates that deep relaxation activates enhanced right-hemisphere activity, associated with spirituality, intuition, sensory output, and auditory/visual awareness. Once the left brain is stilled, full attention may be given to pure, intuitive information. As one enters an even deeper, more relaxing state of regression, continual mind chatter completely stops. At this point, individuals receive "sensory telepathy." Jenny Wade, PhD, conducted a pilot study comparing the similarities and differences between NDEs and post-death regression phenomena. A study participant reported he received "Immediate knowing without all the filters and garbage we put in the way."[52]

Numerous hypnotherapists claim it is nearly impossible for clients to explain past-life telepathy. As past-life regressionist Eric J. Christopher asserts: "Words belong to the realm of our mind, and now information comes from beyond the mind in what can best be described as a 'telepathic download' of simultaneously occurring pictures, sense impressions, insights and/or instantaneous intuitive 'inner knowings.' Clients notice that words feel inadequate to describe what they're experiencing, as if words capture merely the tip of an iceberg."[53]

# Time Distortion

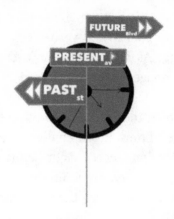

*Image 5-4*
*Representation of Non-linear Time Experienced in*
*Extraordinary Phenomena*
*Time is an Illusion.*[54]
Albert Einstein

In everyday, 3D reality, humanity is governed by time. We attempt to place it in neat and tidy compartments in order to define and organize our lives. Time is measured by schedules and meetings; clocks and calendars; lunar cycles, days, minutes and hours; and the rising and setting of the sun. We categorize it by past (history), present (now), or future (the period after the present moment). The English language is filled with familiar phrases, further entrenching us in the illusion that time is "real." We are all familiar with catchphrases such as:

"Time stood still."

"What time is it?"

"It's time for lunch."

"I had a good time at the party."

It is difficult, if not impossible, for most individuals to imagine a world in which time and space do not exist. However, NDEers, OBEers, UAP contactees, and those who recall past lives consistently describe time as a matrix, or alternate reality, in which the past, present, and future exist simultaneously. Alternatively, time may greatly accelerate or decelerate. Experiencers describe time distortion as:

> Time and space, as we know them, exist only on the Earth realm. When you leave the Earth realm, you leave such constraints. There are realms and dimensions of existence without number, ranging from the slower, denser vibrations of form to higher, finer streams of non-energetic currents. And there is more beyond that, realities that cannot be measured or described in the convenience of mathematics or mind-play.[55]
>
> P.M.H. Atwater,
> Researcher and NDEer

Time and space are modes by which we think and not conditions in which we live.[56]

Albert Einstein

We are inter-dimensional beings. Our minds and consciousness lay beyond our physical brains. Therefore, the experiences we have inter-dimensionally are just as real, because they are real.

Lynn

### NDEs

Dr. Jeffrey Long, radiation oncologist, NDE researcher, and founder of NDERF (Near-Death Experience Research Foundation), conducted a study of over 1,300 NDEers. Nearly 34% of participants reported that during their NDE, time either sped up or everything happened simultaneously, while 60.5% perceived altered time. This sense of distorted time/space is illustrated by one of Dr. Long's respondents:

When I first left my body, I had my diving watch on. I took some very unscientific measurements of the distance I traveled by watching for features and measuring them by the second hand on my watch. Totally unscientific. But my conclusion was and has always been: I was measuring time in an altered time. The ground never moved in a linear fashion; the distances were erratic at best. The distances were always changing ... My intuition and impression were that I was in a different time zone, one where my earthly watch was of no use or inept at making any measurement or reflecting time. Also, without

130

mistake, I would say this whole thing took an hour or more. It seemed to me I was in the NDE for a very long time. But when I asked my diving partners how long I had been unconscious, they estimated five to ten minutes.

Yes, while I was in the light, I had [no] sense of time as I know it here on Earth. In other words, no sense of the serial nature of time ... past, present, or future. All times ... were experienced at every moment in time while I was in the light.[57]

Cardiologist and NDE researcher Pim van Lommel has extensively investigated the correlation between NDEs and concepts from quantum physics. He has concluded that consciousness is not brain-based, thus non-locality is "Everywhere in a dimension that is not tied to time or place, where past, present, and future all exist and are accessible at the same time."[58]

Albert Einstein, the infamous physicist and "father" of the theory of relativity, believed that time, as we know it, is an illusion. In a letter to his family following the loss of his lifelong friend Michele Besso, he wrote: "He [Michele Besso] has departed from this strange world a little ahead of me. That means nothing. People like us, who believe in physics, know that the distinction between past, present, and future is only a stubbornly persistent illusion."[59]

### OBEs

In OBEs, time may either decelerate or accelerate. OBEers can fly, float, or move at the speed of thought. The laws of physics and linear time which apply to waking reality do not apply in the inter-dimensional planes. We can spiral through vortexes of light, travel through wormholes, and enter into vast expanses. Even through the act of moving and walking, one can enter the timeless.

> During an OBE, if I exit my body in my bedroom (this does not always happen) the surroundings look like present time. As I move away from my physical body, away from my bedroom, time becomes more distorted and the environment changes. I am moving into the past and present. Linear time no longer exists.
>
> Lynn

> I hover up, vibrating, and fly again. I'm in whitish space, endless neutral light. I try flying as fast as I can and it's so quick it's impossible to describe—I could go around the circumference of the world in a second at this speed. There's enough room in this white space for absolutely anything and I'm alight with exhilaration. It strikes me that in experiences like this there can be no doubt that we are more than just a physical body. We are physics itself; gravitational pull and light particles and the energy-force that pulls everything together … There's something so harmonious and natural about flying so fast, as if I become the energy of the air itself. There's no resistance and with wonder, I think to myself: "This is soul-flying."[60]
>
> Dr. Clare Johnson

In an article by Montalk, "Astral Physics and Timespace," he describes the physical realm, the etheric realm, and the astral realm in relation to quantum physics and Einstein's theory of general relativity.

First, to understand the quantum physics of the ethereal and astral plane, one needs to start with how we observe our physical reality. We observe our physical reality from the collapse of the "superposition" of waves. What this means is that reality as we know it exists in "waves of possibilities." When we make an observation, these waves collapse into one reality. Multiple outcomes collapse into "one" outcome, and this is what we observe as our reality in linear time. This means that what we see in our physical reality only exists because we observe it. What we perceive as linear time is the collapse of waves into particles. The choices we make will determine the outcome.

The etheric realm is a superset of the physical ... from a quantum viewpoint, the etheric state appears to evolve partial delocalization of the wave function ... In other words, in the etheric realm, the "waves of possibility" do not collapse into a single focal point, and exist all at once. During OBEs we may observe these other realities simultaneously. This implies that within the etheric, one has total freedom to move forward or backward in time just as we physical beings have total freedom to move around in space.

Thus, the etheric is strongly associated with the physical realm, loosely mirroring its shape and diffusing outward in all spatial-temporal directions. It is the seat of raw life-force energy and influencer of probability. In contrast, the astral is as far removed from the etheric as the etheric is from the physical and is thus two orders different from the physical. It is more reflective of internal psychic space than an external physical space.

The astral body is the seat of soul-based emotions. Whereas the etheric pulls on physical quantum events, the astral seems to pull on mental and emotional events. The astral realm, instead of mirroring physical form, symbolically mirrors emotional and psychic energy patterns.

In the Astral planes, the "waves of possibility" do not exist, it is emotional and soul-based; therefore, time does not exist in the astral.[61]

### UAP

The most common form of time distortion that occurs during UAP encounters is missing time. There have been many testimonies by experiencers that witness having seen strange lights in the sky. Afterward, they may see some type of flying object in the form of a spacecraft, or other types of aerial phenomenon. There might be all kinds of strangeness that accompany a sighting such as unusual sounds, temperature changes, changes in electromagnetic frequencies, heightened emotional sensitivity, and even the appearance of animals!

Another common type of time distortion occurs while driving. Suddenly, a driver will see a UAP. Immediately afterwards, the driver has lost track of time, and finds himself on an entirely different road! Minutes or hours may have passed. This is known as "lost time." These occurrences vary greatly from being in your house, walking or hiking, or camping outdoors. Often a person is alone when this happens, but in numerous cases, multiple individuals share the same encounter. Various cases include multiple persons who share the same encounter. Even though sightings seem to occur in secluded areas, they can take place in various locations. The strange lights may be seen by many other witnesses along the same area. Hundreds to thousands of individuals may witness the same phenomena simultaneously.

Many individuals have multiple encounters, and some are lifetime experiencers. Other types of anomalies may follow UAP encounters. Many proclaim missing time during other strange occurrences such as seeing entities whom they communicate with, poltergeist disturbances, hearing strange noises and voices, or having psychic connections where information is shared telepathically.

Brent Raynes, Editor of *Alternate Perceptions* magazine, has close to five decades of experience researching and investigating paranormal events. He has interviewed hundreds of individuals surrounding strange incidents of UAP phenomenon, including missing time. In June of 2011, Raynes interviewed "Allison" regarding her experience with UAP-related contact/missing time. He relates the incident below:

> Allison added that her brother and sister also recalled "being visited by these little people that we thought came out of our bedroom closet." A few years back, her brother and a girlfriend of his were together in a car when they had a pretty close UFO encounter it seemed, followed by missing time. "Neither could recall how they got home," Allison stated. "My brother believes that he is still visited. It really bothers him. It has really affected him emotionally and physically. He's frightened of it."[62]

According to his research, Raynes concluded that certain people may be prone to anomalous perceptions and experiences. They may possess brain chemistry that is different from others. This may include higher concentrations of N-Dimethyltryptamine (DMT), melatonin, magnetite, sensitivity to EM fields, and hemispheric shifts in consciousness. "Whatever is occurring with the UFO contactee/abductee percipient, the human mind itself is the most significant key."[63]

Thus far, our discussion has focused on experiencers who report time speeding up, slowing down, or just plain "missing" during UAP-related phenomena. Yet, the most astonishing form of time distortion is a wormhole, appearing to observers as a tunnel-like structure.

Science defines a wormhole as a hypothetical connection between widely separated regions of space-time. This is an incredibly difficult concept to understand, as it distorts our comprehension that time travels in a linear fashion. Humans perceive time as a three-dimensional construct, and associate it with the past, future, and present. In most basic terms, a wormhole is a tunnel with two open ends, each at a separate location and/or time period. Thus, it is essentially a "shortcut" through space/time, or a portal to another universe. Wormholes are a key element in science fiction because they (theoretically) allow for interstellar, intergalactic, or even inter-universal travel.

Science writer Michelle Starr states, "Wormholes are thought to work a bit like a fold in space. Think of an ant walking across a piece of fabric from one point to another. If the fabric is flat, the ant has to travel the maximum distance. However, if you fold the fabric so the two points almost touch, all the ant has to do is jump from one section to the other."[64]

Barbara witnessed this mind-blowing phenomenon during a UAP-related incident in Miami, Florida.

*Image 5-5*
*Miami, Florida, October 2017, 10:00 p.m.*
*UAP-related wormhole sighting*

In October of 2017, I attended a Consciousness and Contact Conference in Miami, Florida. Prior to the event, a cocktail party was hosted for conference speakers and selected attendees (including myself). Approximately fifty people were in attendance. The event was held at a privately owned home, which, in actuality, was a breath-taking mansion.

One of the attendees (who I will refer to as John) was a self-proclaimed "contact specialist," or one who possesses the ability to initiate UAP contact, via conscious intention. John insisted he could successfully "call down" a craft at will. Alternately fascinated and skeptical of his proposal, guests followed John outside. Heavily bearded and dressed in a turban and caftan-type attire, John impressed me as a kooky woo-woo, a UAP attention-seeker. But what the heck. Phony or not, it sounded like an intriguing adventure.

John stood with his arms outstretched towards the sky. He repeatedly pleaded, "Please make yourself known. I invoke you to show yourself to us." Fifteen minutes passed, yet absolutely nothing happened. What a joke, I thought. I was bored, and nearly ready to call it a night. Indeed, I heard many collective sighs of frustration. At this point, John (who was obviously slightly embarrassed and frustrated himself) requested that we all chant OM (the sound and vibration of the Universe). He believed this would speed up the process of contact. As a final measure, we all agreed to do so. Fifty of us raised our arms to the sky, and along with John, chanted OM. At the time, it seemed pointless; yet, this was the moment that everything shifted, and the inexplicable finally happened.

Let me note that by nature, I am highly sensitive to vibration. I can feel, hear, and occasionally see fluctuating waves of energy. Additionally, I am a Reiki Master, one trained to heal via "universal energy." Reiki training has greatly heightened my pre-existing energetic sensitivities. Yet, what occurred next blew my familiarity with electromagnetic fields out of the water. I have never previously, nor since, experienced anything remotely like it.

Instantaneously, atmospheric conditions changed. The air around me became charged with energy—heavy, dense, and nearly suffocating. Intense vibrations pulsated throughout my body. I was uncomfortable, my body twitching. I knew, without a doubt, that something extraordinary was manifesting (itself). As my body trembled, I saw intermittent, descending flashes of blue and pink lights. They were beautiful, shimmering, tube-like structures, unlike anything I had witnessed on earth.

*Image 5-6*
*Lights in the Sky, Miami, Florida, 2017*

I heard an attendee cry out, "I see a tunnel. It has electricity in it. It looks like a tube to another dimension, an alternate time/space." Another shouted, "I can't see it, but I got a picture on my iPhone. Wow, come see it, it's amazing!" And so I did. I was speechless. The image was jaw-dropping, other-worldly—an actual distortion, or warp of time captured on film.

During the occurrence, I lost awareness of time—it seemed to "freeze." Did the entire incident last seconds or minutes? No one had an answer.

However, we all agreed that the UAP appeared as a wormhole (or traversable wormhole), a tunnel-like structure connecting points that are separated in time/space. In other words, a traversable wormhole is a "space tunnel" linking two distant regions within our universe (or universes). Were the pink and blue structures "beings" from another universe? Perhaps. They certainly appeared to form, then drop from the wormhole.

Regardless, I absolutely believe I witnessed two distinct dimensions of space simultaneously. Time bending, so to speak. This, in itself, is an extreme distortion of time/space, as we understand it.

Wormholes are a common feature of science fiction, not conscious, daily living. How does one find human words to explain such an incomprehensible and mesmerizing occurrence? I could say the experience was jaw-dropping, awe-inspiring, and enormously impactful. However, these words remain woefully inadequate to describe such a magnificent, yet unfathomable incident.

### Past-Life Recall

Reincarnation is not a linear thing. One of the problems in defining it is that there is no analogy to it. It is outside of time, yet we can't help but think of it in terms of the past and the future, and this incarnation. The whole story is so big and involved.[65]

Thomas Sawyer

Bettye B. Binder is the former president of the Association for Past-Life Research and Therapies and has personally regressed over 3,600 individuals. Binder asserts that our true "soul identities" exist in a multi-dimensional universe where time is not limited to a linear construction. As she explains: "Our experience of time is the result of our perceptive mechanisms and basically does not exist outside of three-dimensional reality. It is because the physical senses can only perceive reality a little bit at a time that events seem to exist one moment at a time. We are eternal souls temporarily residing in human form."[66]

Barbara agrees with Binder that past "soul identities," under certain conditions, may be retrieved from an alternate dimension, outside of linear time. As she states: "I have had numerous past-life memories. Perhaps I should refer to them as 'soul identity retrievals.'"

> All of my past life memories have occurred spontaneously. When one occurs, I suddenly "jump" into that previous life, fully immersed in the experience. Instantaneously I am in another body, time, and place—simultaneously residing in two lifetimes and planes of existence. I am completely unaware of time passage while in this state of dual being. At some point, the recall suddenly comes to a halt, and I am fully back in the present. I remember every detail of my experience, yet, am unable to determine its duration. Did it last two seconds? Five minutes? It is often impossible to calculate.
>
> Barbara

Clinical hypnotherapist Dr. Linda Gadbois explains the holographic, multidimensional, and timeless nature of my (and others') past-life memories. As she states:

> Just as we have a subconscious and self-conscious mind on the material plane of the body, our soul, is comprised of both aspects of the mind and has the ability to both be "in the experience" (subconscious), and outside of it witnessing it (self-conscious) from not only a second person perspective, but from a third

person perspective of being able to see it (the current experience or memory) as a part of an ever larger pattern or greater story, in which it's only a part of it. It has the ability to learn from the direct experience being produced by the subconscious, while simultaneously being apart from it and seeing the larger picture as an ongoing story-line that it's merely a part of, or one "act" in a whole series of acts. The soul works according to the Holographic Principle and is conscious and self-aware in both the part and the whole. It experiences itself as both the particle and the wave simultaneously. Present and located in time and the body, while simultaneously existing outside of the body, while "watching" the life of the body as it plays out scene by scene.[67]

It is nearly impossible for humanity to imagine a world in which time does not exist. The structure of modern life demands it. It is a function of earthly reality. Even prehistoric man had a rudimentary understanding of time, which was tracked by sunrise and sunset, seasonal changes, herd migrations, and lunar cycles. Thus, we must keep in mind that time is physically bound. However, extraordinary experiences occur in both altered states of awareness and alternate dimensions. Consequently, extraordinary experiences occur outside of earthly constraints of time, space, physical being, and in multiple dimensions simultaneously.

### Other-Worldly Light

Other-worldly light is also referred to as extraordinary light phenomena (ELP). ELP is atypical light, described as more intense, infinitely brighter, and more radiant than natural light. It appears in numerous forms and shapes. These include non-specific light forms, light balls or beams, auras or halos, flashes of light, light-energy, or an illumination of a place or person. Otherworldly light is generally perceived as golden, white, or blue. Although ELP is most frequently discussed in NDE literature, it is also a common feature in OBEs, UAPs, and past-life regression.

### NDEs

> In every case I have investigated, if the NDEer asked what appeared to be God or a light being or angel if that was what the heavenly host really looked like, the image would immediately dissolve into a burst of radiant light. The individual would then be told that shapes familiar to him or her were used to quell fear and anxiety that the reality of light worlds was beyond human comprehension.
>
> P.M.H. Atwater

One of the core elements of a NDE is rapidly traveling through a tunnel toward(s) a brilliant white, opalescent, or golden light. NDEers describe this light as pure consciousness, a living presence, "God," or "Source." This intense radiance is explained as brighter than one hundred suns, yet not painful to the human eye. It envelopes the experiencer with an overwhelming sense of warmth, peace, indescribable love, and total acceptance. The light communicates profound knowledge to the experiencer. Interpretation of the light appears to be dependent on one's religious beliefs or affiliation. According to Dr. Jeffrey Long's Near-Death Experience Research Foundation (NDERF) study, 64.6% of respondents claimed to have experienced this mystical light.[69] A research study conducted by P.M.H. Atwater revealed:

> 85% of those who answered my questionnaire claimed to have had a scenario where at least half of the episode was filled with

bright, all-consuming light.

52% said they merged into and joined as one with this light (or Being of Light).[70]

A recent study conducted by researchers at the University Hospital of Liège in Belgium concluded that 69% of NDEers encountered a bright, mystical light. Experiencers describe the light phenomena as: "My first visual memory was looking forward and seeing a brilliant bright light, almost like looking directly at the sun. The strange thing was that I could see my feet in front of me, as if I were floating upward in a vertical position. I do not remember passing through a tunnel or anything like that, just floating in the beautiful light. A tremendous amount of warmth and love came from the light."[71]

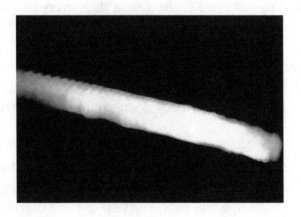

*Image 5-7*
*Being of Light*

"A brilliant white light at the end of a tunnel, and when the wings enveloped me, I became part of the white light."[72]

The light permeated me to my tiniest cell, moving with loving intent toward the deepest

145

parts of me. It loved me and filled me to overflowing.

Penny

"I saw a light that I had never seen on earth. So pure, so intense, so perfect ... At once I realized: there is no time or space here. We're always in the present here. This gave me a profound sense of peace. I felt is as I experienced the light. It's the pinnacle of everything that there is. Of energy, of love, especially, of warmth, of beauty."[73]

## OBEs

Individuals who experience nighttime OBEs describe their surroundings as more brilliant than their everyday, physical environments.

"During my out-of-body experience everything around me was illuminated by a strange glow, like a silvery full moon. When I returned to my body it was 2 am, and the bedroom was dark."[74]

In 2006, Twemlow, Gabbard, and Jones conducted a survey with 339 OBE participants. Thirty percent reported seeing a brilliant white light. Forty-six percent described the light as "extremely attractive," while 33% believed that the light was an actual being.[75]

Layers of energy are often seen as colors of light. The Universe is a multi-dimensional continuum, as we move through these planes of consciousness, movement is also inter-dimensional. Numerous OBEers observe layers of color, which, in actuality, is the mind interpreting the various levels and frequencies of the universe. This correlates to "String Theory," which explains how multiple dimensions are separated by "Membranes." These separations between dimensions are observed during OBEs as colorful membranes of light. I have seen these on several occasions—most often, during experiences when I entered the higher dimensional planes. To be able to witness this amazing manifestation is incredible.

Lynn

*Image 5-8*
*String Theory.*
*Representation of Multidimensional Reality.*

Meeting beings during OBEs is a very common occurrence. Many are reported as spheres of intelligent light, presenting in various colors and shapes.

"I lifted out of my body and floated up to the ceiling. I immediately noticed two globes of blue light in the corner. One of them spoke to my mind with pictures, and I could somehow tell that it was a familiar female presence..."[76]

"I woke up halfway out of my body and saw two large fuzzy balls of light floating in a corner up by the ceiling. I sensed they wanted me to go with them..."[77]

"I had a floating experience when I was nine, and I clearly saw a blue ball of light move around my bed. It hovered above me for several moments and seemed to be examining me. I felt like it was talking to me in my mind."[78]

OBE, May 13, 2018

I turned and looked around at the beauty of this place, the pristine colors of green trees and blue skies. I walked and fell to my knees on the green grass, and I looked up into the sky and said, thank you universe, thank you so much. Pure bliss and love exploded in my heart as I said it. Then a bright spiral yellow light came into view, it was so bright that it blinded me. As I stood up, the bright spiral light was burned into my view; I walked around in a daze, thinking "Wow!" Then I woke up.

Lynn

## Astral Tube - Tunnels of light

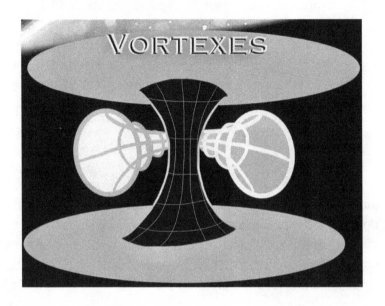

*Image 5-9*
*Representation of vortexes experienced during OBEs, while traveling from one dimension to another*

"As projectors leave their body, they may find themselves traveling through tube-like structures. Many may also find these tubes waiting nearby after exiting the body leaving them a choice to enter the tunnel or not. These astral tubes often look like vortexes and pools of swirling color. Many projectors also describe these as vortexes of swirling light. This light can be extremely bright, colorful, or may appear as blurring stars."[79]

In my personal OBEs, vortexes seem to play a significant role. During ethereal projections (while I am still in my body) I have seen bright colored swirls inches from my face.

According to beings I have encountered, these vortexes are portals into other dimensions and worlds. Once, when I asked to be taken to another world to encounter NHIBs, a vortex opened in front of me. It was so enormous that it filled the entire night sky. It was a beautiful bright blue color, so enormous and spectacular that I just stood there awe-struck, then took a headfirst dive into it.

Lynn

Anecdotal evidence suggests this percentage is higher among people who have experienced OBEs as part of the near-death experience. Unearthly beings have also been reported to give off luminous glows. Also, the astral body is known to naturally radiate light. The term "astral" is derived from the Greek meaning "related to a star."[80]

### UAP-Related Contact

For thousands of years, strange lights have been observed in our skies. Sightings are a worldwide phenomenon that have been viewed by both individual witnesses and hundreds of people simultaneously. Lights from these craft have appeared close to the ground or at high altitudes. UAP navigation lays beyond our current understanding of physics. Contactees insist these lights possess an innate intelligence. Numerous experiencers report lights which intentionally follow their vehicles. Others encounter lights near or above their homes prior to UAP-related contact.

Contactees often report seeing bright light in their rooms or coming from their windows. The difference between unearthly versus everyday light may be attributed to UAPs vibrating at a frequency higher than Earth. If crafts are indeed traveling inter-dimensionally, UAPs may appear in a non-solid form or be perceived as brighter than normal light.

"The experience begins most often when the person is at home in bed and most often at night, though sometimes abductions occur from a car or outdoors. There is an intense blue or white light, a buzzing or humming sound, anxiety or fear, and the sense of an unexplained presence. A craft with flashing lights is seen and the person is transported or 'floated' into it."[81]

Several contactees describe light phenomena as follows:

I went to bed and went to sleep at about eleven p.m. I woke up (and opened my eyes) at about 1:45 a.m. with the perception of my face being bathed in light, which is different than seeing a bright flash. This woke me up totally and I lay there in the dark with my eyes open for a few seconds and then suddenly a bright spot-type light shined in the corner of my bedroom near the closed door.[82]

I was probably about 5 years old or so ... and a bright blue light would come in to the room and the door would open, and there would be like, a foggy kind of misty blue light, just shining through the whole house, and these two figures would come in.[83]

I noticed that the whole house seemed to be lit up with a deep blue light. Turning into the living room, it became obvious that it was emanating in from each window. I put on my shoes to go outside and see who was out there, but by the time I got out there the light had faded away.[84]

There were 2 lights, slightly bigger than the stars zooming around and orbiting each other, making figure-8s etc. And then they stopped moving, got bigger and bigger, until bright light filled the sky and then they were

151

gone. We carried on our journey home. We hadn't stopped for more than 5 mins. The whole journey should have been 20 mins. When we got home, my mum was freaking out, we had been gone for 65 mins, our watches were both still working, but were 45 mins behind every clock in the house. She had neighbors/friends bout looking for us. I don't know what happened to us or what we saw, maybe nothing.[85]

As a kid I lived in a tiny mountain town in Central Oregon. We were very poor, and instead of having a bedroom, I would sleep on a futon in the living room with my younger sisters in our double wide. One night, around the age of 10, I was woken up by the room flooded with a bright white light. At first, I thought a semi must have overturn on 97 and was shining on the curtains. I decided to get up and see what it was, but to my surprise I wasn't able to move at all, much like sleep paralysis. I tried to yell out to my sisters so they could help me, but just as I couldn't move, I found myself unable to speak. At this point I was in a full-on panic attack. Then one of your stereotypical grey aliens passed through the wall and floated me out through the wall to a flight of the navigation shaped craft. The intense light that bathed me made it impossible to see. Suddenly an oxygen mask was put over my face and I blacked out as if on anesthesia.[86]

**Past-Life Recall**
Scientific investigation into light anomalies during past-

life recall is in its infancy. Current research is mainly anecdotal or related to individual accounts. However, this fascinating phenomenon deserves examination.

Carol Bowman is an author, renowned past-life researcher, and founder of the Reincarnation Forum. A forum member describes his experience with light during a self-guided past-life regression: "The imagery I got was very dark, and I felt very alone. I eventually got the impression that I was in space. Video prompted me to look in a mirror, (which was very difficult because in this possible past life, mirrors did not exist.) I finally managed something, and I got a glimpse of myself as an orb, radiating white/yellow light. I could not get a sense of how big or small I was, just a point of consciousness."[87]

A fellow forum member responded to the above post by stating: "I don't know how familiar you are with orbs, but the belief is that an orb is the 'life-force' of an individual that once inhabited a physical body, so I see no reason to believe that it's impossible to visualize yourself as an orb."[88]

As Bowman contends:

An important part of the past life regression session occurs after going through the past life "death," and into a virtual after-life experience. No matter what was recalled in the past life story, going through the death is a distinct shift in consciousness—an altered state within an altered state of consciousness. You experience yourself as a soul consciousness, outside of time and detached from your past or present personality. In this super-conscious state, many surprising things can happen: Some people move into a highly energetic state where they experience healing energies coming into their bodies, imbuing information, love, and understanding. These energies may be experienced as orbs, colors, and other

153

amorphous forms.[89]

Blog writer Bianca Alexander booked a past-life regression session, hoping to experience her own past life (lives). She describes her session as follows:

> With each step, she [the hypnotherapist] guided me to sink into deeper and deeper levels of relaxation, until I fell gently into a hypnotic state. She called in my spirit guides to protect me as I moved beyond this world into the astral realm. Then, she led me gently upward, beyond the forest to 30,000 feet above my body. Floating above myself, below me I saw a chain of iridescent glowing orbs, each representing one of my previous lives. She asked me to choose the one that resonated with me most at the present moment, and led me to gradually float down into that life until I felt myself landing safely on the earth.[90]

Orbs are the beautiful balls of light most readily captured in digital photography and video (note: they can sometimes be seen by the naked eye). They can appear anywhere from wispy to brilliant white spheres of light. These "orbs," with their concentric circles, coronas, mandalas, faces, bumps, knobs and tracks, seemingly appear at random times in one photo, but not the next. They are known to manifest in different colors and sizes, singularly and in groups.

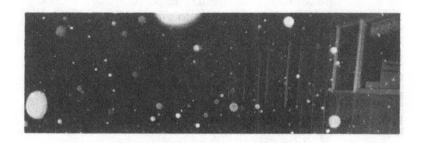

*Image 5-10*
*Photograph of multiple orbs*

Orbs exhibit consciousness and can respond to requests by appearing in meaningful places at meaningful times. Numerous individuals describe the orb phenomena as a spiritual, transcendental, or other worldly experience. Orbs have been sighted during OBEs, NDEs, UAP-related contact, and PLR. Numerous explanations have been suggested to identify these balls of light: soul/ consciousness energy, a vehicle of consciousness, angels, spirits, light waves, discarnate energies, thought forms, inter-dimensional beings, energy fields, aliens, dust or moisture particles, pollution in the atmosphere, or lens flare.

Orbs are mentioned in literature across many genres and have been captured by traditional film, digital photography, and video. There is a growing body of published information, videos, and websites discussing this phenomenon. Though generally not visible to the naked eye, some people can see them while others are able to "feel" their presence.

Our discussion of orbs has focused on those experienced during NDEs, OBEs, UAP-related phenomena, and PLR. However, orbs have additionally been associated with sacred sites, cemeteries, or historic locations. Barbara had a life-altering orb sighting in Stepney Cemetery, Monroe, CT. This occurrence cemented her belief that orbs are a manifestation of intelligent consciousness. Since her experience so closely parallels those reported in the aforementioned modalities, it is detailed below.

Barbara's Orb Sighting
October 16, 1997

My first experience with orbs occurred in Stepney Cemetery, Monroe, CT.

Dating back to the 1700s, the cemetery is steeped in history. French General Comte de Rochambeau played an instrumental role in enabling the Colonists' victory over the British. From June 29–July 1, 1781, Rochambeau's troops camped next to the cemetery, on their way to Yorktown, Virginia, the last major battle of the American Revolution.

Stepney Cemetery is the final resting place of both the late Ed Warren (world-renowned paranormal investigator, demonologist, author, and lecturer), and his wife Lorraine (paranormal investigator, clairvoyant, and light trance medium). Although the Warrens were prolific investigators, the Amityville, NY, investigation, and subsequent book and movie, *The Amityville Horror*, was their most prominent case.

The Warrens occasionally hosted "ghost tours" in Stepney Cemetery, led by their son-in-law Tony Spera (paranormal investigator and current head of the New England Society of Psychic Research). Intrigued by investigating an infamous and haunted graveyard, I signed up for the tour.

Tony had filed a permit with the town of

Monroe to allow our group access to the graveyard from 11p.m.–12 a.m. We filed into the creepy, inky-dark cemetery, filled with excitement and much trepidation.

At midnight Tony announced it was time to go. I was the last person to leave the cemetery. As I neared the exit gate, I gasped. Three blue luminescent orbs of varying sizes instantaneously appeared 100 yards away. The largest orb was approximately five feet in diameter; the smallest, 18 inches. I was grasping a small flash camera; however, I was too overcome with emotion to use it. Thus, and quite unfortunately, I have no photographs of this spectacular sighting.

I am often asked if, at that moment in time, I was afraid. Of course I was. I was terrified. Yet, at the same time, I was utterly mesmerized. The orbs were a beautiful and shimmering, incandescent blue. Many believe that blue orbs symbolize calming and healing energy. They certainly appeared to be benevolent.

The orbs slowly moved towards me in a spectacular manner. They weren't just moving, they were dancing—weaving and bobbing, floating upwards, downwards, diving into the ground and back out. Merging into one another, and then separating, as they slowly approached me—as a moth to a flame.

I sensed that the orbs possessed an innate, deep intelligence—a conscious and spiritual energy. They were moving towards me with purpose and intention. I had the distinct impression they were attracted to my energy, my being, and my ability to perceive them. I stood perfectly still, barely breathing or blinking—never taking my eyes off of this magnificent phenomenon. I eagerly awaited their approach, feeling personal contact was imminent. Yet, Tony shouted, "Barbara, we have to leave now." I was incredibly disappointed.

Over the years, numerous individuals have asked me, "How are you able to connect with and see orbs without a camera?" This is my answer:

Other-worldly entities are believed to vibrate at a higher energetic rate than human beings. I came into this world "wired" differently than most—more attuned to higher energetic frequencies and vibrational levels than others. I believe I can see and communicate with "spirit energy/orbs" because we are a close vibrational "match." Think about it. If orbs are indeed conscious beings, we are not separate from them; they merely reside in a different frequency. If we can dial into this frequency, a "connection" is possible.

I had witnessed an aspect of conscious-ness, unbounded by space and time—materializing in non-human form. It remains one of the most awe-inspiring experiences of my life. I was convinced then, and remain to this day, that I witnessed a physical manifestation of human consciousness.

According to Virginia Hummel, director of TheOrbWhisperer.com:

Orbs come in a variety of colors from the visible light spectrum, including ultra-violet, x-ray and gamma ray range. Currently, there are two different scientific theories as to the physical makeup of orbs; subtle energy i.e. Chi, Prana and Life Force, or plasma.

Orbs are a beautiful demonstration of manifesting physical reality through consciousness. They are the most readily available proof of the existence of life beyond our third dimension. As the veil thins between worlds, we are afforded the opportunity to connect with the magic and beauty of spirit. We have been given a window to the divine. Thanks to the digital camera, we now have a visual experience that quite possibly validates for the first time what mystics, spiritual leaders, and religious doctrine have expressed for millennia: that we truly are eternal beings.[91]

Universally, experiencers describe their encounter with other-worldly light as embracing, comforting, loving, peaceful, blissful, and or healing. For many, it has a life-changing impact, confirming their belief in the continuation of consciousness after physical death.

In the following chapter we examine spontaneous healing. It is an element common to all four modalities; yet, it has traditional medicine scratching its head in disbelief.

# Chapter 6
## It's a Miracle!
## Spontaneous Healing

**Spontaneous healing**

Spontaneous remission or spontaneous regression are both synonyms for spontaneous healing. Regardless of the term used, the definition is the unexpected improvement or complete cure of disease without normal (e.g., mainstream) medical intervention.[1] Many doctors don't believe in the phenomenon (because they were never exposed to it in medical school) and consequently try to discourage people from holding onto what they consider to be false hope and/or believing in false cures. The standard answers the medical establishment gives to spontaneous healings is that there must have been a misdiagnosis or, if any treatment was given (even if known to be ineffective), they attribute this treatment to be the cause of the cure.

Spontaneous remission/regression, also referred to as MR (Miraculous Remission), results in the cure of physical, emotional, and phobic conditions. Miraculous healings indicate that a therapeutic process exists both within and outside of the mind/body. Numerous experiencers maintain that disease, if first arrested on an energetic level, would be unable to manifest on a physical level.

Research and anecdotal evidence demonstrate that spontaneous healing does occur during UAP-related contact, OBEs, past-life regression, and NDEs. Yet, how exactly does this transpire? Alternative healers, Eastern medicine, and numerous experiencers have long contended that disease begins on an energetic level before it occurs on a physical level. This makes sense if we understand that *everything* is comprised of energy. Even thought and emotion have vibrational frequencies. Indeed, we are electromagnetic beings, whose energy extends beyond the physical body, time, and space. Caroline Myss, PhD is renowned in the field of energy medicine. As she explains, "Everything that is alive pulsates with energy and all of this energy contains information ... Even some quantum physicists acknowledge the existence of an electromagnetic field generated by the body's biological processes. Scientists accept that the human body generates electricity because living tissue generates energy."[2]

*Image 6-1*
*Representation of energetic healing*

The late Mellen-Thomas Benedict had a NDE which is generally considered one of the most remarkable ever documented. In 1982, he succumbed to terminal brain cancer. Ninety minutes later, he was resuscitated. Benedict's NDE illustrates the astounding manner in which an energetic, intelligent consciousness is capable of restoring heath:

> I was shown how illnesses start on an energetic level before they become physical. If I chose to go into life, the cancer would be gone from my energy, and my physical body would catch up very quickly. I then understood that when people have medical treatments for

illnesses, it rids the illness only from their body but not from their energy, so the illness returns. I realized if I went back, it would be with a very healthy energy. Then the physical body would catch up to the energetic conditions very quickly and permanently. I was given the understanding that this applies to anything, not only illnesses—physical conditions, psychological conditions, etc. I was "shown" that everything going on in our lives was dependent on this energy around us, created by us. Nothing was solid—we created our surroundings, our conditions, etc. depending where this "energy" was at. The clarity I received around how we get what we do was phenomenal! It's all about where we are energetically. I was made to feel that I was going to see "proof" of this firsthand if I returned back to my body.[3]

### Healing and Near-death Experiences

Inexplicable medical healings have long been reported by NDEers. As the late physician and NDE researcher Dr. Barbara Rommer observed, "I have interviewed many NDEers who re-entered this lifetime cured of the physical illnesses that caused their deaths. I have painstakingly reviewed their complete hospital medical records. My cases include people who have come back totally healed of kidney failure, end-stage liver failure, aplastic anemia (bone marrow shutdown), legal blindness, pneumonia, and cancer."[4]

Dr. Jeffrey Long, radiation oncologist and co-founder of the Near-Death Experience Research Foundation (NDERF), states, "Time and time again, the people who share these cases with the [NDERF] use words like 'miracle,' or 'I was healed.'[5] During the Iraq war, Natalie S. (NDERF survey respondent

6246) was riding in an army truck, when suddenly, a roadside bomb nearly destroyed the vehicle. Natalie, described dead at the scene, was later resuscitated. Grievously injured, Natalie returned from her NDE, relatively unscathed. She later described her experience:

> Having agreed [with the beings] I moved to another vibrational location where healing would be performed on my physical body. From this location, I could see my physical body in the truck, head propped up by my right hand, elbow resting on the door handle exactly as I'd left it. I could also see my body as an energy matrix. Reading from both those levels simultaneously, I could tell that my right hand was nearly severed at the wrist, my right foot and ankle were badly mangled, and I had a deep wound in my right torso. There was a large hole in my head: I was missing one eye, the frontal sinus, and a portion of my brain. Some energy beings and I worked together, quickly repairing the body, primarily working through the matrix. The injuries weren't entirely healed as some were to be of use in situating me for tasks I had agreed to perform, or things I had agreed to experience as a whole infinite self.[6]

"A near-death experiencer, born with cerebral palsy, and a contracted and deformed hand he could not completely open, reported that after his NDE, he was able to open and use his hand for the first time in his life. This medically inexplicable healing was corroborated by his family and health-care team."[7]

I suffered multiple bouts of idiopathic anaphylaxis. Each resulted in respiratory failure, requiring me to be on a ventilator. During my eighteenth bout of anaphylaxis and respiratory arrest, I left my body and traveled to another realm, where I communicated with God. I told Him I didn't want to keep having these attacks. When I asked if there was anything I could do to stop them, He replied, "Say yes when I put opportunities in your path, people for you to help and love, projects to get involved in—say yes." When I woke from my medically induced coma in December of 2016, I decided to give saying "yes" a try. I've said yes to every opportunity God puts in front of me and have been free from anaphylaxis and respiratory failure ever since.

Penny

Anita Moorjani, renowned speaker and New York Times best-selling author of *Dying to be Me*, describes her miraculous healing from end-stage cancer:

At age 42, I found a lump in my shoulder and was diagnosed with cancer of the lymphatic system, better known as lymphoma. For 4 years, my body was ravaged by not only the disease itself but also my fear of it. My weight dropped to 80 pounds. I became too weak to walk on my own. And I started giving up hope that I would heal.

Then, one morning I didn't wake up. My husband rushed me to the hospital where he was told that I had entered a coma and my organs were slowly shutting down, one by one. I didn't have much time left. Prior to this point, doctors had conducted tests on the functionality of my organs, and their report had already been written. But in that realm, the outcome of those tests and the report depended on the decision I had yet to make—whether to live or to continue onward into death. If I chose death, the test results would indicate organ failure. If I chose to come back to physical life, they'd show my organs beginning to function again. About six days after coming out of the ICU, I began to feel a little bit stronger and was starting to walk up and down the hospital corridor for short periods of time before needing to rest.

Every day the doctor reported on my latest test results. "I don't understand. I have scans that show this patient's lymphatic system was ridden with cancer just two weeks ago, but now I can't find a lymph node on her body large enough to even suggest cancer," I heard him say.

To the amazement of the medical team, the arrangements they'd made with the reconstructive surgeon to close the lesions on my neck were unnecessary because the wounds had healed by themselves.

On March 9, 2006, five weeks after entering the hospital, I was released to go home. And I couldn't wait to live my life with joy and abandon![8]

### Healing and OBEs

There are numerous success stories of individuals healing themselves and others during OBEs. The success rate, however, varies. Some may attempt several times with little or no results, while others have had considerable success at achieving one or more of the following:

- A reduction in the severity of the symptoms.
- A rapid healing experience.
- Disappearance of the health issue altogether.

Why are certain individuals adept at self-healing while others are not? Perhaps these individuals *believe* they can heal themselves via focused intention and resolve. This conviction is essential in creating positive outcomes.

The following methods that experiencers have used that have produced success are:

- Symbolically and literally entering and manipulating the energy body.
- Directing healing intent, often manifesting as a type of unexpected light.
- Directing affirmations such as chants or sound energy.
- Creation of symbolic healing imagery.
- Seeking of information about the location, cause, and meaning of the illness.
- Seeking or visualizing a doctor, guide, medicine, or healing environment.

Techniques vary with the use of direct versus indirect, literal versus symbolic, or with varying degrees of internal and external locations of their control. All effective techniques in healing use some form of projective visualization with some being more effective than others.

The success of many experiencers is due to following their intuitional impulses as they make their way through their conscious inner realms. Although they may begin with a general plan or goal, often a spontaneous knowing leads to insights about the situation.

The Buddhists claim that suggestions made during deep meditative states and/or during anomalous experiences such as OBEs are nine times more effective than those made in the waking state. These actions taken "closer to the source" or creativity, and deeper in the subconscious state, increase the healing energies so they may perform much more quickly and profoundly than in the waking state.[9]

Jurgen Ziewe, author of *Multidimensional Man*, writes about his personal journey with OBEs and self-healing.

### OBE and Self-Healing

#### April 12, 2012

When we moved to a new house a few years ago I pulled a muscle in my groin lifting heavy furniture. I was in a lot of pain and found it difficult to move without the pain taking my breath away. On the third day, in the early hours of the morning during meditation, I was able to leave my body and found myself facing it, sitting there in my meditation chair, with the head tilted slightly forward, seemingly asleep. On the left side of my groin I saw a six-inch black hole with dark particles gyrating randomly and chaotically around inside it. I immediately identified it as the trouble spot which had given me so much pain. Not being able to think of anything better to do, I used my out-of-body hands and gently pulled the hole shut, while at the same time sending positive energy into it. Gradually, the hole closed, and the dark energy dissipated. Soon the dark spot had disappeared completely. I then decided to return to my body in order to check whether this was just a hallucination, or whether I had actually closed the "hole" and by doing so, eradicated the problem. When I opened my physical eyes, the pain had completely gone. I got up, moved around, twisted my body, there was not a trace of any pain whatsoever.

Lynn Miller's personal healing experience, via an OBE, on June 23, 2017:

I had gone to Tennessee because my mom was in the hospital. She was in the late stages of Alzheimer's. I had been going through very intense emotions. My mom had almost died, and I was also having to deal with my estranged brother who lives with her. I was making daily trips to the hospital to visit her in ICU. I had gone to my vehicle to get my phone charger. As I was getting into my car (a very high step up into my SUV), I felt a loud pop in the back of my knee. The pain was so intense that I almost passed out. I just sat there. However, I needed to get back into the hospital and I'd parked very far away.

The initial prognosis by my cousin (who is a nurse) and her husband (who is a doctor) was that I had a torn ACL, and would most likely need surgery. I did not have time for that. I was in so much pain that I couldn't walk. I needed to make a trip home to take care of my animals, then back to Tennessee for my mom. I didn't want to take the time or pay for the medical costs of surgery. I am an art teacher (which means finances are always tight). I spend most of the day on my feet walking around my room and couldn't handle the impact of my injury. I felt extreme urgency to try and heal myself. The injury happened on a Friday.

I laid down Sunday around noon before my return to Tennessee, hoping to have an OBE and heal my knee. I asked my guides to help. I felt the best way to heal my knee was to remain in the ethereal plane during my OBE. I remained in my body and did not exit, concentrating on my knee. I saw and was encompassed by a red light, while microscopic fibers of my knee were being pulled together by a loop. After an hour or so, I fell asleep. I was in a lot of pain before my OBE. So much so that I could not walk. When I got up, the pain was almost completely gone, and I could walk. All that remained was a slight stiffness in my knee.

### Healing in UAP-Related Contact

Medical exams are one of the key elements described by contact-experiencers. Thus, it is not surprising that individuals commonly convey medical healings during UAP-related contact. In fact, extraterrestrial healings have been reported by renowned ufology researchers Budd Hopkins, David Jacobs, PhD, and John Mack, MD. Dan Wright, manager of the MUFON (Mutual UFO Network) Abduction Transcription Project, has stated, "That extraterrestrials are curing people should come as no surprise. The central feature to most abduction accounts is the medical examination. This alone should indicate that aliens know a good deal about the human body. But when you add the obvious advanced technology of UFOs, it becomes clear that it is well within the capability of aliens to cure a large number of diseases."[10]

UFO healings have usually been relegated to the realm of the contactee. However, data shows that 45% of cases have occurred to those who have been abducted, as opposed to 28% which involve cases more properly classified as contactees. A surprising 22% of the cases involve people who have had no prior associations with UFOs. Additionally, statistics demonstrate that more men (64%) receive cures than do women (34%). There are also patterns to the way the cures are effected. Twenty-six percent involve surgical operations onboard UFOs. Twenty-one percent involve some type of light beam. Thirteen percent involve alien instruments held over the body, not including the surgical operations. Thirteen percent of the cures occurred apparently as the result of being in close association with a UFO. Nine percent involve pills, salves or medication. Five percent involve injections. Five percent involve healing by alien mind power.[11]

In excerpts from his book *UFO HEALINGS: True Accounts of People Healed by Extraterrestrials*, author Preston Dennett relates several astonishing cases:

Licia Davidson of Los Angeles, California, has been having UFO contacts for as long as she can remember. In 1989, Davidson was diagnosed with terminal cancer. By the time the disease was diagnosed, it had already metastasized to her colon, making it inoperable. She was given three months to live. She then experienced an abduction, during which she was placed on a table and given an extensive operation to cure her cancer. As she says, "I was abducted. They told me I had cancer. They said, 'Relax.' And they did a cure. It was excruciating." Licia was returned to her LA home. Upon her next doctor visit, it was discovered that all traces of her cancer were gone. Licia has recovered her medical records, and states that her cure has been verified by a major medical university. She also states that she fears the United States government (who has harassed her extensively) more than she does the aliens.

This case involves a man by the name of Daniel D. from Harrison County, West Virginia. Daniel reports that he has had many, many contacts with "gray guys" and "tall green, lotus-looking figures." Daniel D. feels that his aliens are quite friendly indeed. As he says, "They've never hurt me, that means a lot ... I think they've helped me in certain ways. I've heard for years that I have to have my wisdom teeth out. One day they were bothering me really bad, and I was in no financial situation to get them fixed, no insurance. That night I was abducted. I can remember, it felt like they were stuffing things on my teeth where my wisdom teeth are. The next day they didn't bother me anymore and haven't bothered me since. That was a couple of months ago."

Although animal healings occur infrequently, they have occasionally been reported.

One summer afternoon in 1959, in Pleasanton, Texas, Susan Nevarez Morton, who was thirteen years old at the time, went to attend a cockfight with her sister and brother-in-law and his family. The entire group drove to the secret location and joined a huge crowd of people. In the center of the crowd, two roosters began to fight a battle which Susan knew would be to the death. The two roosters were fighting fiercely when suddenly Susan's attention was caught by an "enormous red globe with two shafts of white light inching down towards us."

Everyone watched in amazement as the beams targeted the two roosters, who were now lying still on the ground, nearly dead from their injuries. Then something amazing happened, "Their small broken bodies glowed eerily for a few seconds. Then slowly, they both got up on their little chicken feet and began strutting around with robust healthy enthusiasm." The crowd became extremely agitated, until, after a few moments, the beam of light retracted inside the object. Then the object changed from red to orange and streaked away at high speeds. The two roosters showed no trace of their former injuries, but not surprisingly, the cockfight was canceled.[12]

172

The cases discussed above illustrate the necessity for a new paradigm in contemporary medicine—one that is willing to address the question, "Is chronic disease really chronic?" Dennett addresses this in further detail:

> Diseases that we believe chronic are easily cured by extraterrestrials. This alone should reveal that when doctors say a condition is incurable, what they are really saying is that *they* can't cure it. In other words, is there really any such thing as chronic disease? Or does this label conveniently cover our obviously profound lack of knowledge about the human body? In either case, these healing cases have the potential to teach us a great deal about disease, health, and the human organism. If only we could do what the aliens do ...[13]

### Healing in Past-Life Recall/Regression Therapy

PLRT has been described as a method that can offer a glimpse of the mind's enormous capabilities in a variety of ways. This mode of treatment has demonstrated that beneath the levels of consciousness lays profound wisdom, astonishing abilities, and above all, the power to heal one-self and/or empower one to improve life circumstances.[14] A fundamental premise of PLRT is that it operates within a holistic approach, that is, within a paradigm of the mind, body and spirit connection. Often what we have experienced in a previous life is the root of present-day emotional, psychological, or physical issues.

> One way Past-Life Regression Therapy (PLRT) commonly works is that the therapist may simply ask a client to go back to the source of his or her problem. If in a deep enough trance, the client may likely begin to envision himself or herself (in either gender) in a vivid

173

past-life scene that relates to his or her present problem. The image is often so clear, lifelike and vivid that the client feels as though he or she is actually reliving the scene and may express the appropriate emotions as if it were so. After re-visualizing or re-experiencing these images one or more times, and relating it to his or her presenting problem, the client's physical, mental, or emotional symptoms may be greatly reduced, if not completely eliminated. This standard PLT technique is referred to as either catharsis or abreaction.[15]

Psychologist Kevin Hogan further explains abreaction as: "What we have done is to access the memory of the cause of illness or pain, complete with the frame of anxiety, depression, frustration, and other negative emotions, and then reframe the entire memory to one of neutrality, much like a memory of yesterday's breakfast. At this point the illness, pain, or phobia is said to have been abreacted. Abreaction is the discharge of emotion, often grief, anger, resentment, or guilt."[16]

PLRT is most often utilized when symptoms are unresponsive to traditional treatment. Symptoms may be eradicated in as little as one session, whereas conventional therapy may last years. Christopher and Hogan have explained how therapeutic interventions allow patients to retrieve, then discharge negative physical, emotional, or phobic conditions. Yet, how exactly does PLR work on a much deeper, or energetic, level? Perhaps psychological, emotional, or physical diseases are embedded in the energetic fields of our current incarnation. Does past-life regression enable us to let go of negative vibrational patterns carried from one life to another? Individuals under hypnosis appear to make energetic "corrections," alleviating disease and phobias with lightning speed.

Yogic teaching consists of highly sophisticated concepts of both a universal psychic substrate called the *akasha,* which records impressions of all events mental and physical, as well as a vehicle termed *the subtle body,* which transmits individual psychic residues. According to Yogic subtle body theory (the subtle body is one of three bodies that together constitute a human existence), the human body is composed of three distinct levels of subtle energy: the mental, emotional, and etheric bodies.[17]

The mental body is associated with everyday, conscious thought. Psychological issues (i.e., negative thoughts, fears, and phobias) from previous incarnations are embedded in the mental body. The emotional body carries negative emotional residue, including feelings such as anger, resentment, depression, anxiety or jealousy. The etheric body (the energy system of the physical body) is our "physical memory field." The etheric body is imprinted with subtle traces of past-life physical trauma, including diseases, wounds, and chronic illnesses. Simply put, spontaneous healing occurs on physical, psychological, and emotional levels, depending on the type of residual traumatic energy.

*Phobic Healings (imprint on the mental body)*

A phobia is a type of anxiety disorder. It is a strong, irrational fear of something that poses little or no real danger.[18] There are many specific phobias. Among them, the most commonly known are Agoraphobia (fear of public places), Claustrophobia (fear of closed-in places), and Barbara's personal favorite, Arachnophobia (fear of spiders). Specific phobias generally appear in early childhood, around the age of seven. An estimated 9.1% of Americans, more than 19,000,000 people, have a specific irrational fear.[19] Unfortunately, numerous individuals suffer from multiple phobias.

PLRT has been demonstrated to be particularly effective in the eradication of phobias. The manner of one's death in a

175

previous life is intricately tied to a current, overwhelming, and often paralyzing phobia. PLRT discovers the underlying root cause of fear(s) by examining the past life associated with it.

Dr. Brian Weiss, world-renowned psychiatrist and past-life hypnotherapist, recalls a patient's complete elimination of emotional and phobic symptoms after regression therapy. "This was not just the suppression of symptoms nor the gritting of teeth and living through it, a life wracked with fears. This was a cure, the absence of symptoms. And she was radiant, serene, and happy beyond my wildest hopes."[20]

Barbara overcame a crippling fear of knives. This phobia had caused her a lifetime of anxiety.

> From my earliest memory, I have had a terrifying, paralyzing fear of sharp knives. I was obsessed with them—I felt like I had "knife-PTSD/OCD." I had chronically recurring nightmares of being stabbed to death. Even though I knew this was an irrational fear, I had no control over my anxiety. Was I crazy? After all, who in their right mind obsesses about carving utensils? Desperate to get to the root of my phobia, I made an appointment with a past-life regression therapist. Under hypnosis, I recalled a past life in which my father was stabbed to death. The memory of his traumatic death immediately freed me of my fear of knives. I no longer shivered with fear when a Thanksgiving turkey was carved!
>
> Barbara

### Physical Healings (imprint on etheric Body)

As Dr. Samuel Sagan, author and founder of the Clairvision School in Sydney, explains: "I have seen regression bring spectacular results in a vast range of physical disorders, from skin troubles to abdominal tumors, from bladder incontinence to certain forms of paralysis. I have witnessed improvements that amazed fellow physicians or made them pretend nothing had happened, because the results were irreconcilable with their present understanding of disease."[21]

I contracted bronchitis in first grade. This quickly developed into unrelenting, walking pneumonia. My pediatrician prescribed antibiotic after antibiotic, to no avail—he had used up his magic bag of tricks to treat me. No medical intervention alleviated my symptoms. I missed 48 days of school and was just sick of being sick. The principal called my parents in for a meeting, which is nearly always equated with bad news. My parents were told I would be held back a grade if I missed four more days of school. My parents were at their wit's end. "How can Barbara be held back? She's smart, she does her homework, and she behaves in class!" I was terrified. In my mind, only losers were held back. That night I had an exceedingly lucid, realer-than-real dream. I knew I was experiencing my most recent past life.

This particular past life took place in the early 1950s. I was a wealthy British socialite with a two-pack-a-day smoking habit. My physician had delivered devastating news: my years of heavy smoking had left me with terminal lung cancer. The following morning, my six-year-old self awoke. I felt wonderful and knew I was pneumonia free, my lungs clear. My parents rushed me to my pediatrician, who, with a shocked expression, stated: "Barbara's total recovery from pneumonia, literally overnight, is medically inexplicable." Although my parents were stunned, the explanation for my recovery was obvious to me. Recollection of my previous death from lung cancer had spontaneously healed my pneumonia. I had resolved the physical trauma embedded in my mental body.

Barbara

Emotional Healings (imprint on etheric body)

Past-life regression therapist Daniel Olexa has worked extensively with clients who have tried, unsuccessfully, to break negative emotional patterns and limiting beliefs with traditional therapy. Olexa describes a particularly memorable session:

One client in particular was very memorable. As we discussed PLRT as an option, he seemed somewhat spooked by the thought. Interestingly, when I said the word "reincarnation" he perked up. That was the touchstone that resonated for him. This person is an author who was having difficulty promoting his books and believing in himself as an achiever.

As our session evolved, this client visited a past life that took place in the 1920s where he experienced a domineering mother, passive father and a family who did not appreciate his mental/ emotional depth. In this life, his lessons were to keep quiet to maintain the peace and put his desires second to those of others. In short, he didn't matter.

One of the fascinating things to watch when a client is in a past-life memory is their body movement. In the hypnotic state, clients' body language can be significant abreactions. In this case, my client curled up in my recliner, as though he was recoiling from the world that was hurting him in this previous life. We visited the end of this past life; seeing the death scene was freeing to him as he floated up toward space. We talked about the lessons that he learned in this life: being alone, being second, being unappreciated. I asked him if he needed to carry the baggage of these lessons forward into his current life.

He replied, "No."

As I guided him through releasing these faulty beliefs, he sat taller in the chair as each emotional weight was removed. When he woke from the regression a few minutes later, his face was noticeably less strained, he was smiling and energetic. The experience had been profound for him and he took the lessons of the session with him."[22]

We have examined the astonishing, medically inexplicable healings frequently reported by experiencers. However, experiencers maintain that it is the profound, life-altering, and transformational aftereffects which are the most impactful and validating aspect of their spiritual encounters. As P.M.H. Atwater explains: "I invite you to consider the possibility that spiritual development is a physical, tangible process, and that true enlightenment is a vibrational shift in the frequencies of our mental processes. The proof of this, I believe, lies in the after-effects, which clearly demonstrate the degree to which an individual has been able to stabilize and maintain the shift that has occurred."[23]

# Chapter 7
## Transformative After effects

An experience is spiritually transformative when it causes people to perceive themselves and the world profoundly differently: by expanding the individual's identity, augmenting their sensitivities, and thereby altering their values, priorities and appreciation of the purpose of life. This may be triggered by surviving clinical death, or by otherwise sensing an enlarged reality.[1]

NDEs, OBEs, UAP-related contact, and recollection of past lives are life-altering experiences, on spiritual, psychophysical, and social-emotional levels. Terms such as "impactful," "significant," or "otherworldly" are woefully inadequate to describe such phenomena. Indeed, experiencers consistently report extreme difficulty in relating such otherworldly experiences in limiting, human vocabulary. The impact of such events transforms experiencers, permanently altering their personal values, beliefs, conscious awareness, and initiates a larger perspective, spiritual awakening, and a greater understanding of both humanity and the universe itself.

Dr. Brian Weiss is a world-renowned psychiatrist and past-life regression hypnotherapist. The sheer number of dramatic, positive, and healing aftereffects demonstrated by his patients has profoundly changed his own worldview:

> My life has changed drastically. I have become more intuitive. My values and life goals have shifted to a more humanistic, less accumulative focus. I have become more empathic, more loving. I also feel more responsible for my actions. I still write scientific papers ... and lecture at professional meetings. But now I straddle two worlds: the

phenomenal world of the five senses, represented by our bodies and physical needs; and the greater world of the nonphysical planes, represented by our souls and spirits. I [now] know that the worlds are connected, that all is energy.[2]

"The experience changed everything for me: there's something after death, and it's good. Death is merely a release from the body."[3]

"It had such a profound effect on the rest of my life; the timelessness that I experienced; the knowledge that my consciousness will survive outside of my body. It was enough to destabilize my life."[4]

I felt like a package that had been opened, my most important contents removed and reformed, then shoved back into my body, never to fit the same again.

Penny

Through OBEs I have experienced an awakening of my spiritual existence, a transformation of my being. I am more than physical matter; I am more aware and alive. These spiritual experiences disintegrated my original belief system. I no longer believe in the dogma of organized religion and belief systems of society; my knowledge is based upon my experiences. I understand being connected to something greater than myself—to the very source of life itself. To break through barriers of ignorance, fear, and limitation is life-changing.

Lynn

Thus, spiritually transformative experiences

(STEs) are profound and life-altering events. How does one cope with and integrate a sudden, dramatic immersion into an alternate reality?

... There are both positive and negative aspects to the aftereffects ... passing through death's [otherworldly] door seems merely to be "Step One." Integrating the experience is the real adventure—making what was learned real and workable in everyday life. No "set of instructions" covers how to do this.[5]

P.M.H. Atwater

Integration of this ultimately positive transformation may be a difficult, gradual process, lasting years, if not for the duration of one's lifetime. The American Center for the Integration of Spiritually Transformative Experiences (ACISTE) describes the integration process as: "To re-enter one's body or reality after this experience, with a new view of self and life's purpose, much time is needed to process the experience and all of its implications ... Henceforth, experiencers may struggle to integrate or bring into balance the two differing subjective experiences of reality—a physical and non-physical realm, this life and the afterlife, an earthly reality and a spiritual reality." In a recent online ACISTE survey, 12% of respondents reported:

Being able to thoroughly integrate a STE ... it is an ongoing process of growth, work, and learning ... Inhibiting or promoting factors include: How well the sharing of the experience was accepted in significant relationships, the health of the individual, one's life situation, the intensity of the aftereffects, the age of the experiencer, and the manner in which one returned, etc. Common challenges include: processing a radical shift in reality, problems dealing with newly developed psychic abilities, increased sensitivity to electricity, chemicals, smells, sounds, and a yearning to find and live one's purpose.[6]

Let us examine the major transformative aspects of each experience in addition to the integration process itself.

## Transformational aftereffects
### NDEs
#### Increase

- Concern with spiritual matters 42% [7]
- More spiritual/less religious 80–99% [8]
- More generous and charitable 80–99% [9]
- A desire to help others 73% [10]
- Compassion for others 57.3–73% [11]
- Appreciation for ordinary things in life 84% [12]
- Concern with nature/environment 84% [13]
- Understanding what life is about 63% [14]
- Ability to love others 73% [15]
- More open and accepting 80–99% [16]

#### Decrease

- Concern with material things in life decreases 50–79% [17]
- Interest in organized religion decreases 80–99% [18]
- Fear of death decreases 80–99% [19]

### Psychophysical aftereffects:

- More sensitive to other realities 50–79% [20]
- Became more psychic 80–99% [21]
- Became more telepathic 86% [22]
- Immune system strengthens 80–95% [23]
- Healing ability 50–79% [24]
- Increased awareness of spirits 65% [25]
- Precognitive abilities increased 50–79% [26]

- Enhanced intuition 92% [27]
- Increase/change in energy level 80–95% [28]
- Heightened response to taste 50–79% [29]

### Assimilation

Transformative aftereffects tend to be presented in positive terms. However, the assimilation of such profound changes into everyday life is often painful, confusing, and slow to process. Research indicates this adjustment period, on average, takes seven years. Why is this process so challenging? As leading researcher and three-time NDEer P.M.H. Atwater explains:

> The experience and its aftereffects can be a challenge to live with. The urge to serve, the depth of compassion and empathy experiencers come to display, the desire to "walk with God," the extent to which unconditional love begins to influence everyday life routines—all of this does not shield near-death experiencers ... from depression, confusion, or disorientation. Some report no problems whatsoever in adjusting to "life as always," but the majority I have had sessions with find that they must face and deal with some very difficult issues.[30]

One of the greatest challenges NDEers face is the impact of their experience on family members, who struggle to understand their loved one's experience, question its validity, and are perplexed by the "stranger" their loved one has become. Separation becomes common-place. Approximately 75% of NDEers divorce within seven to ten years. Spouses commonly report they no longer know or understand their loved one. P.M.H. Atwater found: "The general mindset was that significant others were convinced that the experiencer was out-of-touch with reality, while the experiencer becomes convinced that significant others were slow to move forward and were not interested in making changes. It was as if the two ... started speaking different languages and could no longer communicate effectively." [31]

My relationship with my fiancé, Don, became increasingly strained. The more I searched for answers and meaning to help me assimilate my experience, the further apart he and I became. It felt like there was no one I could talk to. No one understood what I'd been through and how terribly homesick I was for Heaven. I longed to go back to the light of God. I tried talking to my doctor and was met with a blank stare.

Don felt I was losing touch with reality and worried he was in a relationship with a woman he no longer knew. I felt his understanding of life and eternity was stunted and wondered if we'd ever find common ground. I began researching near-death experiences, and the challenges people face following a NDE, and discovered 75% of experiencers end up divorced. Finding the statistic troubling, I showed it to my fiancé. I looked at him and said, "Surely the fact that I've stood in the presence of God shouldn't hurt our relationship. Do we want to be in the 75% or can we work together and get through this?"

It was Dr. Atwater's research that made us look at my experience, my new "self," and our relationship. Was this something we could sort through and come out stronger on the other side? The answer was a resounding "yes." Almost four years later, I can tell you that we have worked through my NDE and the aftereffects, and we are both happier and stronger for having stuck with it.

Penny

Therapy is frequently needed to help the NDEer move forward. However, it is challenging to find a therapist qualified to handle this type of experience. Counseling is most effective when the therapist is also an experiencer. Marriage and family therapy may also be necessary when changes in the experiencer are incongruent with his/her partner.

## Transformative Side Effects
## OBEs

> *Perhaps the most important benefit received from out-of-body experiences is the recognition of our personal ability to discover the answers for ourselves.*[32]
>
> William Buhlman

## Impact of OBEs
### *Immediately after experience*

- Became interested in psychic phenomena 78% [33]
- Felt curious 68% [34]
- Felt one possessed psychic abilities 40% [35]
- Felt life was changed 55% [36]
- Developed a greater sense of awareness 83% [37]
- Felt experience had lasting benefits 71% [38]
- Heightened belief in life after death 63% [39]
- Believed OBE was a spiritual experience 51% [40]
- More peaceful and harmonious relationships 54% [41]
- Heightened interest in knowledge and science [42]
- More tolerant and understanding attitudes 44% [43]

- Decreased interest in materialism 50% [44]

### *A longer time after the experience*

- Developed a greater awareness of reality 83% [45]
- Felt experience had lasting benefit 71% [46]
- Felt a change toward a belief in life after death 63% [47]
- Felt experience had great beauty 61% [48]
- Felt experience was the greatest thing that ever happened 40% [49]
- Felt experience was reminiscent of childhood experience 20% [50]

### *Assimilation*

The transformative effects of OBEs are generally reported as positive; however, assimilation and adjustment are quite different than NDEs. First, OBEs do not involve a life and death situation, which is a profound component of NDEs. Some may have spontaneous OBEs throughout their lives, so these incidents are not as dramatic. Those who practice controlled OBEs are often on a type of spiritual journey within themselves (lasting years or even decades). The biggest difficulty is awakening to the aspect of how their inner world is at times very different than the physical world. Some who experience or practice OBEs do not fit society's norms and report feeling alienated from peers and family.

Although numerous OBEers seek counseling, traditional psychotherapy is not an effective treatment modality. Psychologists and psychiatrists tend to diagnose experiencers as pathological or abnormal. Research indicates that transpersonal therapy (a holistic therapy focusing on the social-emotional, creative, intellectual needs of a patient, with an emphasis on spiritual health) facilitates healing and integration. Unlike conventional therapy, a transpersonal counselor is trained to respond to spiritual crises with open-mindedness. For example, a therapist may prompt a reluctant experiencer to open up by asking:

> *Therapist*: I would like to ask you a question that may seem a little strange at first, but I think your response may be relevant to what we have been discussing about your dream experiences. I wonder, in the past, have you ever experienced yourself separate from your physical body, as though for a moment your sense of consciousness departed from your body and traveled somewhere else? If yes, what was that experience like for you? [51]

Dr. Stuart Twemlow emphasizes the significance of a therapeutic model which views OBEs as non-pathological and spiritually transformative experiences. As he explains:

"The therapist should take on the role of a teacher, encouraging clients to learn more about the nature of OBEs and come to a greater understanding of what their OBE might mean in the grand scheme of their life."[52] It is vital for counselors to consider how OBEs impact a client's relationships, perception about life and death (from an existential therapy framework, for instance), or how the experience affected the client's connection with the spiritual dimension of life.[53] Talk therapy is highly successful when combined with meditation, biofeedback, self-suggestion relaxation techniques, and/or dream work.

**UAP-related Contactees**

*Transformative Aftereffects*

*Psychological*
- 72% become more sensitive/intuitive [54]
- 79% become more psychic [55]

- 50% (for at least a short period) are able to heal others by touch [56]
- 57% reported paranormal activity began after contact [57]
- Anecdotal evidence suggests concern with spiritual matters strongly increases, interest in organized religion strongly decreases
- Evidence suggest strongly decreased fear of death

**Psychophysical-Omega Project by Kenneth Ring** [58]

- Need less sleep
- 15% overwhelmed with more info downloads than they can handle
- 70% have increase in allergies
- Developed Electrical Sensitivity Syndrome
- Increased sensitivity to light
- Increased hearing acuity
- Increased fluctuation in mood

### *Integration*

UAP-related contact experiencers (CErs) undergo a journey of integration quite like that of NDEers. This occurs in many stages, and not always in linear order. Many CErs refuse to discuss their experience, assuming others, including family members, will label them as mentally ill or deluded. They may deny the occurrence altogether, maintain social isolation, and struggle with anxiety. Like NDEers, CErs frequently seek emotional support in the form of counseling or hypnotherapy. Experiencers often join online or organizational support groups. Both NDEers and UAP-related contact experiencers may suffer depression as they attempt to reconcile the experience with their spiritual beliefs or previously held values and lifestyles.

It is only when CErs begin to truly examine their experience that transformation can occur. As clinician Roberta Colasanti, former clinical director of PEER (Program for

Exceptional Experience Research) explains:

> Once people investigate this, they will often say that it is as though they have permanently crossed a threshold. The reality they once accepted without question has been irrevocably changed. Once they have decided that this is indeed happening in their lives and they are having these experiences, the world as they knew it, ceases to be ... this stage is about what meaning these experiences have for people, and in what reality they hold these experiences ... Acceptance is a later stage. At this stage, a person will say something to the effect of, "I don't care anymore what the culture says about these experiences; this is happening to me." In other words, external validation is no longer needed.[59]

Life-altering positive transformation begins at this stage. Numerous CErs describe their transformation in mystical terms. The descriptions are nearly indistinguishable from those of NDEers. Both speak of a newfound, awe-inspiring connection to humanity, life and God, Source, or Creator.

### Past-life recall

*Psychophysical Aftereffects Increased:*
- More sensitive to other realities: Awareness of "all" being connected via energy [60]
- Ability to heal: 23% [61]
- Reported incidence of patient healing: 50% emotional/psycho-physical/phobic [62]

*Psychological*
- Loss of fear of death [63]
- Greater sense of purpose [64]

- More peaceful and content [65]
- Shift towards more transcendent beliefs [66]
- Calmer, less anxious [67]
- Greater sense of interconnectedness with others and the universe [68]
- Happier, more satisfied [69]
- View life from a more positive and meaningful perspective [70]
- Religious views have broadened [71]
- Better self-esteem and self-image [72]

The beneficial outcomes of past-life experiences are numerous. Healing benefits include loss of fear of death, a greater sense of interconnectedness, enhanced compassion, and a more positive view of life. Additionally, experiencers report feeling greater peace/contentment, happier and more satisfied with life, a larger sense of identity, less alone and connected to the universe, and emotional, psychological, and/or phobic healing. A study conducted by Dr. Heather Friedman Rivera, JD, PhD, and clinical hypnotherapist determined, "The major influences that affect the degree of benefit from past-life experiences are the number of experiences, the vividness and what prompted the experience. However, the number of experiences is the primary influencer."[73]

Each of my past life memories have taught me valuable and often profound lessons. Yes, several have produced spontaneous physical and phobic healings. Others, however, have been spiritually and emotionally trans-formational. I'm talking life-altering trans-formation on a soul/ energetic/DNA level. It's so difficult to put into words. Human vocabulary is utterly limiting in describing transcendent phenomena that defies human conception of space/time. I recall at least two lifetimes in which I was callous, judgmental, indifferent, and yes, even cruel towards others. The memories of my heartless behavior have deeply affected me. In my current incarnation, I feel physically ill if a violent scene occurs on a television show. I become highly anxious if I see an individual being treated unkindly and am obsessed with justice. I am lucky these past-life memories occurred when I was a child, so that my worldview was transformed early in life. I have always realized that as interconnected beings, we must treat one another with kindness, respect, and without judgment.

Barbara

I saw myself in past lives where I consistently stood up for the oppressed and mistreated. In each life I suffered a martyr's death. It became clear to me that I was not required to live lives where I consistently put my own needs on the back burner, that it was okay for me to make myself a priority. God didn't require this of me. It changed my life because, for the first time, I realized the importance of loving and caring for myself. I learned that God wants us to cherish our own lives and live them to the fullest, and that we are not prescribed a lifetime of guilt for doing so. In searching my soul, I saw that while I routinely got involved in standing up for others, I secretly resented the misfortune it brought on myself and my family. I live differently now that I have this knowledge. I love taking care of my own needs and those of my family. I don't jump headfirst into the problems of others and realize I don't have an obligation to always be involved in "fixing" the problems of others. I live a more balanced life now. I help when and where I can, but I also honor that each person grows through overcoming his personal adversities.

Penny

Past-life memories often impact family dynamics. This is especially prevalent when the experiencer is a child. Children who remember past lives will frequently make comments such as:

"I have another mommy and daddy," or "I have different parents."

"When I was big, I ... (used to have blue eyes/had a car, etc.)."

"That happened before I was in mommy's tummy."

"I have a wife/husband/children."

"I used to ... (drive a truck/live in another town, etc.)."

"I died ... (in a car accident/after I fell, etc.)."

"Remember when I ... (lived in that other house/was your daddy, etc.)."[74]

How does a parent react to such remarks? We are aware that our children have vivid imaginations and engage in creative play. Most of us encourage this. Imaginary playmates? No problem—we barely break a sweat. What happens, however, when our child repeatedly insists, "I was a chimney sweep in England. I had a wife and five children. We were very poor. I died after I fell off a roof in 1905." Do we shake our heads in disbelief? Insist our child is fantasizing or perhaps emotionally unstable?

Most parents are confused—unsure how to react or

respond.

As one mother explains:

> When Ryan was four, he began directing imaginary movies. Shouts of "Action!" often echoed from his room. But the play became a concern for [us] when he began waking up in the middle of the night screaming and clutching his chest, saying he dreamed his heart exploded when he was in Hollywood. His mother asked his doctor about the episodes. Night terrors, the doctor said. He'll outgrow them. Then one night, [as I] tucked Ryan into bed, Ryan suddenly took hold of [my] hand. "Mama," he said. "I think I used to be someone else."
>
> He said he remembered a big white house and a swimming pool. It was in Hollywood, many miles from his Oklahoma home. He said he had three sons, but that he couldn't remember their names. He began to cry, asking [me] over and over why he couldn't remember their names. "I really didn't know what to do," [I] said. "I was more in shock than anything. He was so insistent about it. After that night, he kept talking about it, kept getting upset about not being able to remember those names. I started researching the Internet about reincarnation. I even got some books from the library on Hollywood, thinking their pictures might help him. I didn't tell anyone for months."[75]

Research indicates that children who recall past lives are emotionally stable. In fact, parents of such children are often more upset than the child describing his experience.[76] Dr. Jim B. Tucker, Professor of

Psychiatry and Neurobehavioral Sciences at the University of Virginia School of Medicine, further suggests:

> We recommend that parents be open to what their children are reporting. Some of the children show a lot of emotional intensity regarding these issues, and parents should be respectful in listening just as they are with other subjects that their children bring up. When a child talks about a past life, we suggest that parents avoid asking a lot of pointed questions. This could be upsetting to the child and, more importantly from our standpoint, could lead the child to make up answers to the questions. It would then be difficult or impossible to separate memories from fantasy. If children persist in saying they want their old family or old home, it might be helpful to explain that while they may have had another family in a previous life, their current family is the one they have for this life. Parents should acknowledge and value what their children have told them while making clear that the past life is truly in the past. Hearing a child describe the experience of dying in a painful or difficult way can be hard, but both parent and child can know that the child is safe now in this life.[77]

# Electromagnetic Aftereffects

*Image 7-1*
*Blocking EM Waves*

Electromagnetic effects (EMEs) are a particularly fascinating aftereffect of OBEs, NDEs, UAP-related contact, and in some cases, PLR. It has been suggested that electromagnetic sensitivity validates the mystical belief that spirit/soul is comprised of energy. Consciousness is a form of EM energy, and with focused intention it can achieve extraordinary things. Human perception occurs due to the interactions between the subatomic particles of our brains and the quantum energy sea. We literally resonate with the cosmos.[78]

In *Thunderbolts of the Gods*, Wallace Thornhill and David Talbott describe an electric universe in which: "From the smallest particle to the largest galactic formation, a web of electrical circuitry connects and unifies all of nature, organizing galaxies, energizing stars, giving birth to planets and, on our own world, controlling weather and animating biological organisms. There are no isolated islands in an electric universe."[79]

EME aftereffects include interference/malfunction in a myriad of electronics, including computers; lights; light bulbs; watches; vacuum cleaners; insulin pumps; defibrillators; electric toothbrushes; and GPS devices, among others. Additional similarities between EM after-effects in both NDEs and UAP-related contact include tingling; headache; fatigue; low blood pressure; post-traumatic stress disorder (PTSD) (40% reported after NDEs); depression; and inexplicable healing. According to P.M.H. Atwater, nearly three-quarters, or 73% of NDERs have extreme sensitivity to atmospheric changes, such as thunderstorms, earthquakes, lightning, and tornadoes; and additionally, report feeling "drained" when in proximity to electronic devices.[80]

### NDEs

P.M.H. Atwater is a three-time NDEer. Atwater describes

how her now off-the-chart electromagnetic sensitivities have impacted her public engagements:

> During the 2001 IANDS Conference held in Seattle, I was the first speaker on the docket before Melvin Morse, MD. Soon after I began, the overhead lights started to undulate off and on in waves and patterns that increased in intensity. Everyone in the room saw it ... I was so embarrassed. I stopped talking and prayed with all my heart that they stop. Finally, they did. No one else experienced anything like that, nor did the lights "misbehave" again. During the 2003 IANDS Conference in Hawaii, I was seated about seven to eight feet from the recording equipment used for Bruce Greyson's keynote address. Afterward, the tape was blank. Bruce's entire presentation was forever lost. I was "asked" by IANDS after that to never again sit close to recording equipment at their conferences. During a presentation I was giving about the aftereffects of near-death states to the Psychology Class at James Madison University in Harrisonburg, Virginia, I mentioned electrical sensitivity, and the lights popped from a sudden power surge. This only affected the building where I was. This type of thing has happened so often at talks I have given, that I have learned to calm my energy down before I speak. Sometimes that works, and sometimes it doesn't.[81]

Numerous NDEers claim they "vibrate at a higher frequency" and have been "rewired," or "re-aligned" since their experience. An experiencer reported to researchers Greyson, Liester, Kinsey, and Alsum, "I awakened. I now see all things as energy in motion and all connected. After dying, my mind began to understand the energetic route in and out of my body, and I'm now free to open and close those energetic doorways."[82]

Perhaps the white light (commonly described by NDEers as pure, loving energy) "re-wires" an individual's energy field, resulting in heightened electro-sensitivity afterwards. It appears suggestive that energetic re-wiring may also occur in those individuals experiencing contact with NHIBs, who as an advanced species, could potentially vibrate at a higher energetic frequency than earthly beings.

It is widely accepted that NDEers (and additionally, as indicated above, OBEers, UAP-related CErs, and past-life experiencers) share core transformative aftereffects, demonstrated by the data collected from numerous researchers.

### OBEs

A common symptom of OBEs are energetic surges and tingling in various body areas. It's as if the body starts to become rewired into a higher electro-magnetic field. These energetic sensations are different from the OBE vibrational state, mainly because they can be felt for hours or up to days following an OBE. When one is experiencing a higher state of consciousness during an OBE experience, one's energetic vibrational frequency also rises. Our journeys into the depths of our consciousness open the gates of unresolved emotional trauma, anxieties, and fears in our daily lives. When I was first beginning my controlled OBE experiences, I often faced my fears. I either conquered them or learned to accept and love them. Fears manifest themselves in several ways and are often symbolic. Our emotional blocks can come in the form of walls to break down or scary creatures to fight. What an amazing gift that OBEs offer us, to break through these barriers of our psyche. It is very liberating. The energy sensations can be felt in specific areas of the body depending on the type of emotional blocks we are experiencing. The energy is correlated to our chakras and the emotional and physical areas of the body they correlate to. Liberating these energies enables us to heal both physically and emotionally.

Lynn

### The Vibrational State

One of the most commonly reported phenomena associated with out-of-body experiences is the vibrational state (VS). In a survey conducted by William Buhlman with over 16,000 respondents, 85% experienced sounds such as buzzing, humming, or roaring, and 56% experienced vibrations or high energy sensations.[83]

The vibrational state transcends culture, race, sex, religious views, age, nationality, and level of education. Not only is it experienced as a very tangible phenomena that is powerfully felt throughout the entire body, it is also the most audible sensation associated with OBEs. Vibrations may sound like a freight train or jet engine is taking off in your head or may be very subtle.

The VS is a prelude to the OBE and is often associated with another common sensation known as sleep paralysis. Not all individuals experience the VS prior to an OBE, and many may have the VS without having and OBE.

The VS can be quite shocking and terrifying if you do not understand what is happening. When my OBEs became a frequent occurrence, I began to worry that I might have something wrong with my brain, like an aneurysm or stroke! I began to research this phenomenon to better understand it and quell my fears. Now I welcome the VS because I know great adventures into consciousness follow them.

Lynn

I felt vibrations and a noise travel up from by belly area into my head. Loud buzzing, vibrational sound. I wanted to go with it, but at the time I was shocked/ scared ... The vibrations started, and they were intense. I felt as my whole body was being shocked with electricity, the way I described it at the time was like lying on a bed of corrugated iron but that the corrugations were moving rapidly down the length of my body. This shocked me back into full wakefulness. I remember just lying there and thinking "what the hell just happened to me?" It took me a long time to fall asleep after that because I was afraid that the vibrations might happen again.[84]

Excerpt from William Buhlman, Astralinfo.org

The out-of-body experience may raise the energetic frequency of our bodies. An OBE transcends space and time. It is a journey into consciousness. As one moves from the lower frequencies of the dense physical body into higher vibrational frequencies, energetic changes occur. Energy blocks are then cleared to encompass a wider, universal energy. Vitality and energy will increase in the physical body, enabling emotional and physical healing to occur.

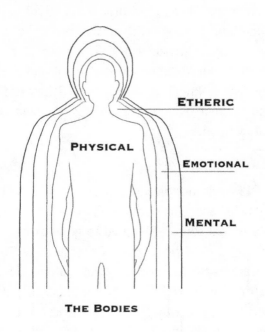

THE BODIES

*Image 7-2*
*Subtle Energetic Bodies*

When consciousness shifts inter-dimensionally, there is a change in the physical EMF. Following separation, the physical body pulls on the energy body. The closer the proximity of the two bodies, the greater the magnetic force. This can make the OBE experience end quickly, as the physical body pulls its consciousness back in. It feels like an overstretched rubber band, popping back into place. As the energy body moves away, the magnetism decreases. That is why many will move immediately away from the physical body, enabling more freedom. This connection will never break. At any moment we can go back to our bodies. This connection has been referred to as "the silver cord."

**UAP-related Contact**

Richard Bonenfant is a retired research scientist and lecturer, who offers insight into heightened EME fields common to both NDEers and abductees. He has stated:

> It is evident that both of these groups share a common crisis of cognition, specifically the realization that reality extends beyond the limits of what is currently believed to be possible. To experience the continuity of consciousness following physical death or to be faced with the reality of being abducted by aliens should certainly fit this criterion.

He has additionally explored the possibility that EM effects are caused by the interplay of bio-physical energy and external EM forces.[85]

Experiencers report a variety of EM escalation following contact/abductions:

> "I moved to the city and was renting a house with my daughter. The light bulbs in the house would literally blow over my head showering me [with] bits of glass. The electrician couldn't explain it." Another respondent noted frequent difficulties and/or computer malfunctions and feeling "jinxed" around electronic devices. She additionally experienced lights flickering or turning off and on by themselves or blowing out in her presence.[86]

A study conducted by Kathleen Marden and Denise Stoner (2012) determined that 32% of UAP-contactees (non-abductees) and 68% of abductees reported problems with electronic equipment.

Several interviewees noted:

> Malfunctioning computers, watches, appliances, radios, televisions, cameras and compasses.

> Some reported that light bulbs blow out or blink off and on and the hands on their watches spin.

> One participant stated that the time clock at his place of employment malfunctions only for him.

> Others mentioned that streetlights blink out when they drive or walk under them. Another stated that when she became emotionally upset light bulbs would "shoot in the room I was standing in." Televisions, appliances and computers would automatically switch themselves off and on.[87]

### Electromagnetic fields in PLR

How is one's electromagnetic field heightened after past-life regression? As aforementioned, unresolved physical or emotional traumas from past lives are embedded in the energetic fields of our current physical body. Every thought and emotion vibrates at a certain electromagnetic frequency. Thus, if trauma from a previous life remain unaddressed, our energetic bodies remain "bogged down," vibrating at a dense, or lower frequency. Consequently, trauma resolved from a past-life regression therapy or spontaneous past-life memory elevates our energetic frequency.

As Reverend Edwige Bingue explains:

> When you're lower in frequency, you're lower in consciousness—your perception is dulled—and you view things through a foggy lens. When you're higher in frequency, you're

higher in consciousness, your perception is heightened, and you see things more clearly the higher you go. Just as when you ascend in an airplane, you see a larger picture the higher you climb, and the brighter it becomes as you break through the clouds ... Vibrational frequency depends on the amount of Life Force you are channeling through you. Life Force can also be characterized as "God Energy and Intelligence." The difference between the manifestations of the physical, mental, emotional and spiritual result simply from different levels of vibrating energy, or frequencies.[88]

Formal research on the correlation between PLR and EME aftereffects is in its infancy. However, anecdotal evidence suggests that past-life regression therapy increases conscious awareness, empathy, psychic ability, and electromagnetic sensitivities.

I experienced my first past-life memory when I was six years old. Language is woefully inadequate to describe the enormous impact it had on my life. Shortly after this occurrence, I began to notice a "shift" in my awareness. My psychic abilities grew. I just "knew things" yet couldn't not explain how I did.

For example, I developed an uncanny psychic ability to "read the minds" of family members. I almost always knew what my parents were going to say before the words came out of their mouths. This did not go over well. I always answered them before they had finished asking me a question. They were both baffled, yet extremely annoyed by this newfound capability. I was constantly being told I was "rude," which earned me the nickname "the interrupter."

I experienced frequent paranormal occurrences. I became deeply empathic. My parents began calling me "over-sensitive" in a derogatory manner. I was told I was being dramatic and sought attention, because "no one is *that* sensitive." I could feel electromagnetic fields around me. I felt as if my entire being had been physiologically and energetically rewired. This was quite overwhelming at times, as I was merely a child.

Barbara

We have examined the transformative aftereffects of extraordinary experiences. The authors have shared their own wondrous and life-changing journeys. Now, let us examine six renowned individuals whose lives have been forever changed via their own spiritual encounters. These individuals include experts in the fields of NDEs, OBEs, paranormal phenomena, ufology, psychotherapy, past-life regression therapy, and neuroscience. Now, let Erica McKenzie, Bob Peterson, Brent Raynes, Marcie Klevens, Heather Friedman Rivera, and Bob Davis take us on their own journeys of personal transformation.

# Chapter 8
## Weighing In-Our Stories

*"We all have that divine moment, when our lives are transformed by the knowledge of the truth."*[1]
Lailah Gifty Akita, Inspirational Writer

*"The world is full of magic things, patiently waiting for our senses to grow sharper."*[2]
W.B. Yeats, Poet

*"The unexamined life is not worth living."*[3]
Socrates, Greek Philosopher

### Near-Death Experience, October 2, 2002

#### Erica's story

During my first nursing job I was introduced to two of the most widely prescribed drugs/drug combinations for weight loss and weight control. Memories and feelings of the ridicule and bullying I experienced in the past negatively impacted how I viewed myself. It seemed everyone was still concerned with how I looked, and I couldn't get away from the pressure to be thin. These messages were everywhere; in the tabloids and magazine covers in the grocery checkout line, on television, and even in the sideways glances from strangers. The pressure of society's expectations of what was an acceptable physical appearance was too much, and once again, I found myself trying to fit in.

At 125 pounds, I made the decision to try these drugs with a goal of losing ten pounds. I had a lifelong history of partaking in abusive behaviors in hopes of improving my physical appearance. I had been bulimic for twelve years and had gotten to the point that binging and purging my food, combined with running four miles a day, wasn't enough to reach my goal weight.

One of the most serious side effects linked to these diet drugs is primary pulmonary hypertension, or PPH, a serious lung disease in which blood vessels constrict and create abnormally high blood pressure. Unfortunately, PPH is usually a permanent condition, and in many instances can lead to death. Other side effects of PPH include shortness of breath, chest pain, swelling of extremities, palpitations, and heart valve leakage. Valve damage is generally a silent condition, causing no symptoms to the patient, until the condition becomes severe. In addition to these cardio-pulmonary complications, many individuals also experience neuro-ransmitter deficiencies.

People with neurotransmitter deficiencies may suffer changes in brain function, an inability to concentrate, learning disorders, anxiety, panic attacks, PMS, menopausal symptoms, obesity, impulsive behavior, fatigue, fibromyalgia, low libido, insomnia, short-term memory loss, headaches, irritability, personality changes, and depression.

Dieting and eating disorders are the most common cause of self-induced neurotransmitter deficiencies. In addition, long-term use of diet pills can deplete neuro-transmitter stores. Diet pills like Phen-fen and Phentermine use up large amounts of dopamine and serotonin, which can result in a multitude of health issues. Prolonged emotional or physical stress, abnormal sleep, and inflammatory conditions like endometriosis can all compound the severity of neurotransmitter deficiencies.

In 1997, several users of Fen-Phen and Redux reported many of these side effects and resulting damage from taking these drugs. As a result, lawsuits and class action suits were filed against the drug manufacturers and eventually, the FDA pulled these drugs from the market for a period of time.

Unfortunately, I was still able to get access to these drugs and continued to take Fen-Phen for close to nine years. Over time, I started sensing something was wrong with my physical and mental health. I began to exhibit several of the side effects. I slept and ate very little, had manic episodes of highs and lows, and exhibited behavior that resembled bipolar disorder. Undeterred by these symptoms, I chose to ignore my health concerns because I was addicted to the drugs.

I was alone and away from home when suddenly, my symptoms took an extreme turn for the worse. I was having trouble catching my breath and felt as though I was going to pass out. I encountered a stranger who drove me to his church just up the hill in hopes of getting me help. Within minutes, he pulled into the parking lot where a pastor was waiting to help him. I didn't have the strength to walk, so the two men assisted me inside the church. Unbeknownst to me, the pastor had already dialed 911. The pastor looked to me and asked, "Child, who are you?" I couldn't answer. I couldn't even remember my name.

"Are you married? Do you have a family?" I had absolutely no idea. I couldn't recall anything. I tried to draw a breath but couldn't. I commanded my lungs to open and fill with air, but they wouldn't. I leapt from my chair and jumped up and down to force air into my lungs, but this time it didn't work.

In the distance, I could hear the paramedics coming down the hall. They might as well have been miles away because, in that moment, I knew it was too late. They could not save me.

*"Help me!"* I wanted to scream, but I had no breath to utter words. I was terrified. I was suffocating. *I knew at that very moment; I was going to die.* The pastor was a strong man of substantial size. He sprung from his chair and pounded his fist on the desk. "Speak to me now because I can see it in your eyes. I know where you're going." His voice was fading. "Child! Do you believe in God?"

With the last ounce of energy I had in my body, I cried out, "I believe in God!" As I said those words, my soul separated from my physical body, and I watched from above as my body collapsed to the floor below me.

Those were the last moments of my human existence on this planet seventeen years ago. What followed was a near-death experience so remarkable that each time I allow myself to replay this experience, I feel as though it just happened to me yesterday. Its lessons continue to unfold.

As I took my last breath, the most extraordinary thing happened. I was instantly filled with unconditional love—beyond human description. For the first time in my life, I could really "breathe!" There was no pain as I drifted away from my earthly body. Yet, I didn't want to leave my current life. Suddenly, an angel appeared just above me and to my right. This angelic presence reassured me that everything would be fine, if I just *let go*. I let go, and sensed the physical mass, which had once contained "me" was now instantly gone—replaced with an exhilarating consciousness. At first, there was only a deep, inky darkness. Suddenly, a tunnel appeared, filled with the most brilliant and exceedingly powerful light. This light was permeated with an overwhelming sense of intoxicating love. The acceleration with which I traveled through the tunnel was supersonic, yet effortless.

I traveled further into the all-encompassing light until I finally reached the end of the tunnel. Once there, an incredibly intense feeling of love overwhelmed me. Human words cannot accurately describe the enormity of the love that filled each cell in my body. This love was so tangible I could touch it. While I could not see a physical image, I distinguished this brilliant presence as "The Voice" I had heard all my life. It was God. God embraced me. We stood together with our backs towards Heaven and looked towards the stars. Light radiated all around me. The most powerful love emanated from this other-worldly place. I knew I was finally home. I never wanted to be separated from that feeling or from God again.

God and I began to communicate telepathically. Although

He shared numerous lessons with me, I would like to focus on one in particular, "The Rippling Effect." God told me to look to my right. Suddenly, an arm and shoulder appeared in human form. They were the size of a semi-truck. I watched the arm extend forward and then swiftly, swing straight up. It extended to the furthest stars until it was no longer visible. Suddenly, the hand reappeared. Resting in its palm was a massive rock. It was bigger than the largest boulder I had ever seen on Earth. Emanating from this rock was the most brilliant light.

God turned to me and said, "You are the rock. You are the Light. The Light is of Me and I am with you." Immediately afterwards, God let go of the rock, and together we watched it fall for what felt like a lifetime. I repeated to myself, "I am the Light. I am the rock." I saw a vast body of water appear. It was greater than the largest ocean and appeared endless. Into the water the rock plunged. I could feel the great force of its impact. Together, we watched as a single ripple of water appeared. God said, "Mankind is the water. You are the ripple." I watched the single ripple expand until I could no longer see it. I repeated, "I am the ripple." And then God turned to me and said, "Like the ripple affects the water, so too do man's words and actions affect mankind. You are the rock. You are the light. You are the ripple that affects mankind."

I began to comprehend the significance of these words. Each word, thought, and action, no matter how small, affected everyone and everything. I had no idea of the power and consequences of my thoughts, words, and actions. I knew it was imperative that I remember this if I returned to Earth. I had it all wrong. When we come into this world, we are all the rock. The light dwells in each of us if we allow it. This light comes from the one who created us. We have a choice to let that light shine or not. The stronger our connection to God, the brighter the light shines within us.

Reflecting on the rock as it made the water ripple, I came

to understand that we too will leave our imprint, making an impact on a level most of us can't begin to comprehend.

I could no longer focus on my insignificance and imperfections. I could see that I had to begin making a valuable impact on the Earthly plane, by learning to love myself the way God loved me. No one had the qualifications necessary to define me or to take my power. I felt so much peace knowing that although I am just one person, this love and light means that I can make a difference by carefully choosing my thoughts, words, and actions. Fueled by love, I understood that if we come together with this individual power we carry within, the ripple would be unstoppable. It could change everything.

I hadn't considered the possibility of going back to my physical body until God imparted the following: Until you learn to love yourself on the Earthly plane, you won't be unable to fulfill your mission; to heal and grow the gifts you have been given.

Leaving Planet Heaven, I traveled down a dark tunnel, making a pit stop at the edge of Hell. While there, I had a handful of experiences before being returned to the tunnel. Eventually, I made my way back to my lifeless physical body.

Dying taught me so many lessons and has impacted me on a divine spiritual, emotional, mental, physiological, and cellular level. I'm not the person that I was prior to having this experience. I've realized my challenging life experiences, flaws, imperfections, mistakes and past choices were not mistakes or roadblocks, but rather, opportunities for learning. If I choose to learn from these, I have the ability to heal and awaken. This is also the case for everyone. I've come to understand how important it is to look to God, rather than mankind, for my value. Our uniqueness is our value. Our value is our contribution on this Earthly Journey.

Since my NDE, I have working hard to heal and have become a self-advocate for my spiritual, mental, and physical health. I have been inspired to combine my NDE and medical knowledge, dedicating myself to researching NDEs and non-ordinary phenomenon. I'm determined to educate the public regarding such experiences by supporting and sharing other people's stories. I choose to view these stories as learning opportunities for others. Perhaps, this is a necessary "medicine," required to create healing and divine connection.

### Erica's Bio:

Erica McKenzie, BSN, RN, hid a lifelong battle with addiction and body image until it ended nearly 17 years ago, when she collapsed to the floor unconscious, and had an extraordinary near-death experience (NDE). She is now dedicated to encouraging people who have had a near-death, or some other non-ordinary occurrence, to communicate about their own experiences. This dedication recently gained the attention of Roland Griffiths, PhD, Head of Johns Hopkins University School of Medicine States of Consciousness Research Team. He extended her an invitation to serve as the ambassador for their new anonymous, web-based, global survey study regarding near-death or other non-ordinary experiences. Erica has been labeled one of the up and coming inspirational "new thought leaders," with a recent guest appearance on *The Dr. Oz Show*. She is the author of the book *Dying to Fit In.*

### Out-of-Body Experience

### Bob Peterson's story:

I've had a successful career in software engineering—frequently ranked top among my peers. From an early age, I've always been logical and analytical. While other kids were outside playing games like baseball, I was usually reading science books. When I saw my first computer in 1976, it was love at first sight. I ditched the science books and started reading computer tech manuals for entertainment. I was shallow and materialistic, eager to start my career, earn some money, and buy that American Dream.

I was also raised Catholic. Although I thought of myself as scientific, I held deeply rooted religious beliefs based on the Bible as the word of God; such as original sin, salvation through Jesus, and so on. I blindly followed the Bible's code of ethics. Like most science-oriented people, I did not see a conflict between scientific and religious beliefs; they were simply placed in two separate boxes. This was the foundation of my entire, closed-minded belief system.

That all changed in September 1979, when I was eighteen and read *Journeys Out of the Body*, by Robert Monroe. Monroe had all these absurd stories about leaving his body, flying around like a ghost, and coming back to reanimate his body again. However, as laughable as this sounded, Monroe didn't actually ask anyone to believe him. Instead, he said, "Try it for yourself and see." He even provided a method to induce an out-of-body experience (OBE).

A large part of a scientist's job is to formulate a theory based on experience, design experiments, test those theories, and then publish the results. Monroe claimed he could leave his body, so my theory was that he was either lying, dreaming, or hallucinating. I decided to test his ridiculous claims by trying his procedure myself. I could certainly spend a few hours to assume the role of lab rat. "What have I got to lose?" I thought.

I carefully followed the book's instructions to the letter. After about an hour, just when I was about to give up, I felt a very strange sensation, like a "twang" inside my head. Then, a sudden rush of weird electrical energy swept into my body. Suddenly, I literally felt like I was being electrocuted. Through my closed eyelids, I could see a bright blue ring of electrical fire rushing straight for my head. Instinctively, I tried to raise my arms to brace for impact, but found, to my horror, that I was paralyzed. I heard my heart pounding in unbridled fear, but I was helpless; unable to move. After thrashing and kicking wildly out of sheer panic, I finally regained the use of my limbs, and the paralysis wore off. That was my first encounter with "the vibrations," a common precursor to an OBE.

My life was suddenly turned upside down. I had come very close to an out-of-body experience. My mind reeled with the implications. There was no room in physics for the non-physical, so science must be wrong. However, I now understood that science, not Monroe, had been lying all these years. But that wasn't all. There was no room in my Catholic religion for out-of-body states either; Catholicism was lying too. So, if I couldn't trust what science or religion had taught me, what, or whom could I believe? Everybody in my life was ignorant about this. Surely my computer buddies would laugh hysterically at the idea. I only knew one person I could trust: me. I knew that, from that point on, the only thing I could trust was my own personal experience. Seeing is believing.

From that point on I had a new mission: To find the "Truth" with a capital T. As much as my brief encounter had scared me half to death, I needed to learn how to induce OBEs more reliably, and test Monroe's other claims. Suddenly, there was a whole new world to explore, and surely, Monroe had only scratched the surface. I felt like Ferdinand Magellan or Christopher Columbus—sailing on a monster-filled ocean toward the unknown edge of the Earth, afraid of falling off, but driven by the need to explore. I made a sacred vow to myself to collect and read every OBE book available, until I knew everything about the subject. I pledged to try every OBE technique until I could do it at will. I promised to endure the fear to get to the truth and swore to share what I learned with the world.

Before my encounter with the vibrations, I blindly followed science. They had somehow duped me to believe scientific dogma, physics, and brain physiology. I no longer

215

accepted that limiting set of beliefs. There was a whole new non-physical world to explore, one they completely missed.

Similarly, before my encounter, I also blindly followed my religion. Catholicism had somehow duped me into believing in the "natural" and "supernatural" as separate things, rather than one homogeneous whole. Suddenly, it was clear to me that the stories in the Bible were based on other's experiences of the same non-physical world I had just brushed upon. However, these individuals had been dead for thousands of years. Their messages were distorted not only by lack of words to convey what had happened, but by thousands of years of cultural changes and language misinterpretations. Even after 1500 years, people believed all kinds of unscientific nonsense. Giordano Bruno had been burned at the stake for saying the Earth revolved around the sun. This did not just apply to Catholicism—all the world's religions were guilty. It made me wonder—why would billions of people believe all those cockamamie stories while the same religious experiences are available to everyone today?

I am now in my mid-fifties and have honored the sacred vows I made to myself to the best of my ability. My OBE library now has more than 200 volumes and is growing. This does not include my books on peripheral topics like near-death experiences, shamanism, and other altered states of consciousness. I did learn to induce OBEs; many hundreds of them and continue to do so. The delicious otherworldly feeling never gets old. I have shared what I have learned with the world. I have written four books and am currently working on my fifth.

Like all the scientists and explorers who came before me, my quest is never-ending, and my journey continues. The more I learn, the more I realize my ignorance. I have learned to value the journey itself. My greatest hope is to open minds, expand consciousness, and change people's lives, as Monroe changed mine.

**Bob's Bio:**

Robert (Bob) Peterson has been studying and inducing out-of-body experiences and psychic experiences since high school. He graduated from the University of Minnesota Institute of Technology with a BS in Computer Science. While attending the University, he became proficient at out-of-body exploration and kept detailed journals of his experiences. Bob volunteered for a student-based organization called the Minnesota Society for Parapsychological Research (MSPR), which gave him experience as a "ghost-buster" before the movie made the subject popular. While living in Phoenix, Arizona, he began editing and contributing articles for a local, spiritually-themed newsletter, "The Spontaneous Self." Bob has had a very successful career in systems level computer programming. While working as a contract programmer for IBM, he compiled his experiences and journals into his first book, *Out of Body Experiences: How to Have Them and What to Expect*. Bob has written four books and is currently at work on his fifth. He lives near Brainerd, Minnesota, with his wife and soulmate, Kathy. Bob's website is:
http://www.robertpeterson.org

### UAP-Related Contact

**Brent's Story:**

Allow me to introduce my humble self. I am but another human soul, struggling to make my way through the trials and tribulations of everyday life, on this tiny, spinning speck in a vast universe. In fact, it is so vast, that I can't even begin to truly imagine the tremendous depth and dimension of this ginormous realm.

As a child, I was blessed, as many children are, with a pretty fertile imagination. Awhile back, I did an interview for my online magazine, Alternate Perceptions (ap-magazine.info) with the noted California psychologist Dr. Jon Klimo. Jon, like myself, reflected on his early interest in the great mysteries of life, including UFOs and the paranormal. His interest was initially sparked in childhood, while gazing up into the starry heavens above in awe and wonderment. In my youth—as I'm sure many, many others have done too—I looked up with awe and wonder into the starry firmament above. I read books on astronomy, the future space travel plans of NASA, and watched science fiction flicks including *Forbidden Planet* and *The Day the Earth Stood Still.* I understood, even then, that there was more going on here (and, of course, out there) than what I could perceive in my daily home and school life. Deciding to become what I later learned was termed a "ufologist" at age 14 (a title that despite no college education, I was able to jump right onboard), my initial perspective of the matter was quite simplistic. I was inspired by noted Indiana writer and broadcaster Frank Edwards. His book, entitled *Flying Saucers—Serious Business,* contained many intriguing stories that came from seemingly credible individuals worldwide.

In the beginning, it seemed obvious that the popular mainstream ufological emphasis on the "nuts and bolts" extraterrestrial theory was the only logical way to go. It seemed clear to me (as well as the Air Force, numerous astronomers, armchair skeptics, and psychologists) that such phenomena were the result of misinterpretations of natural phenomena, delusional people, and/or hoaxes. Perhaps ET visitations did occur. However, my sole focus on that either/or proposition was quickly challenged by the writings of three popular UFO authors: Brad Steiger, Dr. Jacques Vallee, and the one and only John A. Keel. All put forth alternative theories, ideas, and evidence that examined ufology, in many instances, in a rather dramatically different light.

By 1969, my perspective on the UFO enigma had undergone a significant shift. The writings of New York journalist and noted UFO author John Keel, without question, had the greatest impact on this shift. When I was seventeen, I began corresponding with him. I sought his advice and insights on how to initiate my own field investigations. I had been very impressed with his magazine articles that described his field work in various states; in particular, West Virginia, Ohio, and Long Island. "Essentially the contactee experience is identical to religious apparition phenomenon and probably is caused by the same factors," Keel wrote me. "It might be best to familiarize yourself with the medical and psychiatric studies of the religious cases before you tackle the UFO variation."

Although no college education was needed to adopt the honorary title of "ufologist," I was a teenager, and Keel's research suggestions sounded quite academic. Well actually, it still does. Nonetheless, I did my best to adhere to Keel's advice. I corresponded and met with mental health professionals who showed any interest in the UFO subject. I also met and corresponded extensively (for well over three decades) with the late psychiatrist Berthold E. Schwarz. Dr. Schwarz had become a close friend of Keel's after an article he wrote, entitled "UFOs: Delusion or Dilemma?" had been published in *Medical Times* (October 1968, No. 10). The article discussed four UFO cases. In each, Schwarz had personally interviewed the main witnesses. Schwarz had intended this to be his swan song on the UFO subject, but instead, the journal feature came to Keel's attention. Shortly afterwards, the two were engaged in extensive correspondence. At this point, Schwarz had learned from Keel that the "paranormal" played a highly significant role in the lives of many UFO experiencers. Schwarz had already spent years researching, investigating, and writing about para-psychological phenomena and events. Thus, including UFO data in his studies was easily accomplished and highly impactful. Schwarz learned from both Keel and his own field investigations that poltergeist phenomena, psychic events, and manifestations were frequently described by UFO experiencers as occurring soon after their close encounters and contact episodes. Additionally, experiencers frequently had a lifetime of such episodes, prior to any remembered UFO encounter event.

A woman in Florida reported a close encounter with the classic domed disc-shaped craft. Within days after her sighting, poltergeist activity and apparitional phenomena began in her home. She put me in touch with Schwarz soon after I met her in 1973. In 1975, I spent nearly the entire summer traveling from Maine to Florida, meeting and interviewing many UFO researchers and experiencers. I tried to get a handle on this matter. I came across much UFO data with a psychic interface. In late October, I returned to my home state of Maine, and became involved in a reported close encounter case of two young men. Their encounter was followed by—as you might have guessed—reported poltergeist type activity. I immediately got Schwarz on the phone and briefed him on the situation. Within three months, Schwarz was in Maine, talking with the young witnesses and their parents.

In 1976, Schwarz wrote an article describing his investigation into this case in England's well recognized journal, *The Flying Saucer Review*. The journal also included an article by me and another by a fellow researcher, Shirley C. Fickett. In 1983, Schwarz had a massive two volume tome published, entitled *UFO Dynamics*. It included my article on the Maine case, in addition to Shirley Fickett's. Schwarz's book covered many cases from various states. During my extensive UFO-related travels in the 1970s, I followed up, with the good doctor's help, on a number of interesting situations that he had been involved with.

A husband-and-wife UFO investigative team in Fairfield, Ohio, often allowed me to rest my weary head on a bed in their basement. They had some pretty interesting UFO experiences themselves. The wife described seeing a silver-suited humanoid figure who would occasionally appear and disappear in her home. Years later, one of her adult daughters shared her memory of seeing this mysterious figure when she was a child. Additionally, from time to time, the family claimed unusual psychic, poltergeist-type manifestations.

Again and again, if you're out in the field investigating these stories, you will find an unusually high number of these high-strangeness accounts from witnesses who, for all intentions and purposes, seem like ordinary, everyday, and trustworthy folks. Keel made a strong case in his writings back in the late 1960s and early 1970s, describing all of the seemingly interrelated, yet overlooked UFO cases where paranormal occurrences, apparitions, and cryptids showed up; yet for years (and still today), many "ufologists" would not include such details in their reports, because they see them as unrelated to their particular field. For example, most parapsychologists, cryptozoologists, and "ghost hunters" do not wish to venture outside of the self-imposed boundary lines of their presumed disciplines.

Too many individuals are too focused on the deceptively misleading surface appearance of these anomalous reports. Instead, attention should be directed to conducting an objective and thorough comparative analysis of anomalous data to determine if these phenomena may indeed be interrelated.

UFO author Tim Beckley interviewed John Keel, congratulating him on his commitment to investigate the interconnection of UAP and paranormal phenomena. Keel's theory was accepted by numerous ufologists worldwide. "So you've triumphed in the end," Beckley said to Keel. "How does it feel?" "It's a hollow victory," Keel said. "We have just opened Pandora's box. Instead of solving the mystery, we've created many new ones."

The Pandora's box Keel alluded to is a huge and complex challenge. It demands a multidisciplinary focus, discerning, and objective attitude towards the data. Keel wrote to me stating that one had to "peel away layers and layers of nonsense to get at the 'truth.'" Sadly, ufology, which clamors so much about getting the government to disclose the truth about UFOs, has long hidden the vital truth from itself, over and over again.

We've often been our own worst enemy. This needs to change.

### Brent's Bio:

Brent Raynes has been immersed in ufology since 1967. He is the editor of the monthly Alternate Perceptions online magazine (apmagazine.info). Brent is the author of *Visitors from Hidden Realms* (2004), *On the Edge of Reality* (2009), and *John A. Keel: The Man, The Myths, and the Ongoing Mysteries* (2019). He is a contributing author to *Beyond UFOs: The Science of Consciousness and Contact with Non-Human Intelligence (2018)*. Raynes has lectured extensively throughout the country, and has appeared on Coast to Coast AM radio, Art Bell's Midnight in The Desert radio, The Paracast, and Exploring the Bizarre, in addition to hosting his own audio programs for Alternate Perceptions. Brent investigates reports of close encounters, contact experiences, and the parapsychological and psycho-spiritual components of those occurrences.

Brent Raynes may be contacted at:
brentraynes@yahoo.com.

### Psychotherapy for Paranormal Experiencers

### Marcie's story:

There are many roads that lead us to spiritual transformation. The road I took was off the beaten path, to say the least. I am a psychotherapist in private practice in Seattle, Washington, and I'm also a lifelong experiencer of the contact phenomenon. My journey began at the age of two when I encountered a strange being in my living room one night, and it has twisted and turned throughout my life. I have seen how these experiences can ultimately lead to a much broader view of reality. I spent the first half of my life trying to deny the reality of the double life I led, and the second, working to figure out how to integrate it and overcome my fear. The phenomenon facilitated a shift in my consciousness and the way I perceive the world. Over the years, I've learned how to integrate these two realities and finally become whole. It has taken me many years and a lot of help to get here.

I've been asked to write about the ways that I work with experiencers of the paranormal in my therapy practice. To start, I must begin by talking about many of the issues that myself and others have struggled with, in trying to come to grips with paranormal phenomena. I struggled with the reality of the phenomena in my life for many years, wanting to believe in anything, except what was actually happening to me. I grew up in a very logically based family; my first job was working in my father's law firm. In the world I came from, it wasn't real until it was notarized on paper, and this phenomenon didn't fit into my world. Experiencers of the contact phenomenon are faced with two choices: you can either choose to doubt your own perceptions or loose the security of your worldview. To accept the reality of what was happening to me, I would also have to allow myself to fall into a state of total ontological shock.

I've spent many years researching and attempting to grapple with experiences that left me feeling lost and alone, with a growing mountain of questions. I struggled to find my footing in a culture that denies the reality of experiences that transformed my life in profound ways. I've seen firsthand the amount of stress that this type of secret can bring, and how hard it is to function on a daily basis under its weight. I have seen and experienced how these phenomena can stress and even destroy relationships with loved ones. I know the feeling of helplessness when you realize that your children may be involved and you're helpless to stop it. These are just a few of the types of crises that experiencers face on a daily basis. When you add isolation to the mix, you can sometimes end up with depression, PTSD, low self-esteem, and anxiety.

It took me many years to finally allow the utter meltdown of the way I viewed reality. I went kicking and screaming most of the way. The ability to regain your footing and recover once your worldview has been smashed is one of the hardest tasks that lay in the path of all paranormal experiencers. Probably the deepest and most fundamental injury one can have to their psyche is whether you can trust your own perceptions of reality. When you're faced with this type of a choice, it is easy to see how many people often fall into dissociation. Dissociating or blocking the memories out can be a valuable defense mechanism at certain times in your life. Don't judge yourself harshly if you are not in a place in your life where you feel supported enough to delve into your experiences. It's important to carefully assess whether you are in a stable place in your life, with support around you, before you decide to delve further into your experiences.

I made the decision to become a licensed psychotherapist in hopes of providing more support for people experiencing these types of phenomena. I am very familiar with the crises these types of experiences can bring and I am also equally aware of how spiritually transformative they can become. As a therapist, I like to work with people in a multi-pronged approach. In my work with experiencers, I use several different modalities, including eye movement desensitization and reprocessing (EMDR), hypnotherapy, and spiritual emergence or emergency coaching. These approaches enable me to work with people on many different levels at once.

EMDR is one of the first tools I use to help experiencers to process and often overcome the fear they may have around their experiences. EMDR can help to desensitize a lot of the pain and fear that can accompany some anomalous experiences and help people to remember more detail. It's definitely one of the best tools for PTSD recovery that I have found. EMDR can be used to install new, more empowering messages into the subconscious layers of the mind. These new messages can help people to remain more lucid during the experiences or help them to remember to ask certain questions during future encounters. If you find that you are suffering with symptoms of PTSD, such as insomnia, depression, nightmares, flashbacks, or hyper vigilance in your environment, then it is important to get help and find a therapist trained in EMDR.

The next tool I often use is hypnotherapy. It is a wonderful tool for healing and can also help people to recover more detail in their memories. In using hypnotherapy with my clients, I have found that people often experience relief, enhanced understanding, less confusion, and more awareness of the different ways the phenomenon is operating in their life. Techniques can be used during the relaxed state of hypnosis to relieve pressure and aid in healing and soul retrieval, if necessary. These are techniques that can start to desensitize and help you to release the fear you are holding in your body.

The last tool I will discuss is spiritual emergency coaching. Spiritual emergency is a transformational crisis that can cause a person to become overwhelmed by the experiences of non-ordinary states of consciousness. These states can sometimes be accompanied by confusion or preoccupation and distancing of loved ones, to the point it may become hard to function in daily life. These experiences can be triggered by spiritual practices like yoga and meditation, or other factors like near-death experience, contact phenomena, fasting, trauma, grief, recreational drugs, etc. I help my clients with grounding techniques and psycho-education to empower them to move through the different phases of a crisis more smoothly. These tools also help them to discover the deeper learning these experiences often bring. People often emerge from the crisis much stronger, wiser, and more open—often able to discover their life's true path and follow it.

As I look back on my own integration process, I realize that I vacillated between periods of dissociation, denial, grief, fear, and feelings of awe. It took me many years until I finally reached acceptance. Integration may take many years to achieve and has been an ongoing process in my life. I found that it was extremely important for me to begin by grounding the experiences into my life by taking physical action. Physical actions like researching, journaling, drawing pictures, joining a support group, connecting with others in night watches, or becoming a field investigator can be very helpful. When journaling about your experiences, it's important to include dates, times, and feelings you were having before and after the experience. It can help you to accept the reality of the experiences when you read about others having experienced the same things, about which you had written years earlier.

I have seen people grapple with the inability to accept that the experiences they have had are actually real. This state of denial can be an impediment, which slows the process of healing. For years I struggled with dissociation and doubt about the reality of what I was experiencing. It was at that point that I made the decision to turn to the philosopher René Descartes for inspiration. "I think, therefore I am" being one of his most famous quotes. René Descartes tapped into the concept that, at the most basic level, we are all "consciousness."

When I embraced the idea that I am consciousness, I realized that these experiences had touched me on the deepest level of my soul, profoundly changing me on many levels. I realized that from this level of consciousness, I couldn't think of anything that was more real than these experiences that altered who I am in profound ways. This expansion of my worldview allowed me to conceptualize the phenomena into my life in a new way. It helped me to break through and finally own what had happened to me. The moment I owned it was when the healing process could finally begin.

### Marcie's Bio:

Marcie is a psychotherapist in Seattle, who specializes in working with paranormal experiencers. She earned her master's degree in clinical psychology from Saybrook University. Marcie is trained in both hypnotherapy and EMDR, which are both outstanding tools to help people to process and move through trauma. She understands the transformative nature of non-ordinary experiences because she's a lifelong experiencer herself. With determination and help, she was able to move through the fear and into empowerment; it's her mission to help others do the same. She facilitates the Washington State Para-normal/UFO Encounter Support Group and believes in the need for community through these experiences. She is also a psychological consultant for Experiencer Research Team (ERT) of MUFON. Marcie is currently in private practice in Seattle at Open Box Counseling.

### Past-life Regression Therapy

### Heather's Story:

In 1997, I was diagnosed with rheumatoid arthritis and then ten years later, fibromyalgia jumped on board. I must have signed up for the bonus package because a few years ago, I lost part of my hearing and now experience tinnitus or ringing in the ears constantly. Hearing aids help to some extent and also serve to help me handle the unnerving constant ringing.

When first diagnosed with rheumatoid arthritis, I followed conventional Western medicine, but it was not enough. I sought to take an active role in my health and wellbeing. I could not sit around for the three months between

227

appointments waiting for the doctor to tell me what to do next. To be honest, I was terrified.

As a registered nurse, I had seen the devastating effects of this disease on patients. I was a go-getter, working hard, raising children and rarely taking time for self-care. I figured I would do that when I was old. And then, at the age of 33, two days after a hysterectomy, I found I could not turn on the shower faucet. My first thought was that somehow, I had broken my hand. But how could that be when all I had been doing was laying around recuperating from surgery?

A few doctors and tests later, we had a diagnosis . . . rheumatoid arthritis. I must have looked at the lab results twenty times, thinking I misread it. "Yup . . . It says rheumatoid factor-positive . . . Wait . . . let me check again . . . hmm, still positive."

I am not sure how long it took my brain to become convinced that this was now my reality. I felt as if I just received a life sentence. Maybe I am being a bit dramatic, but that is my nature. I have a habit of pushing an idea to the absurd; to make it a catastrophe. I guess it is just my way of dealing with the negative. Then I realize it is not as bad as I imagined and come back to some middle ground.

I would come to understand the meaning of the word "chronic." My lifestyle, up to this point, was not conducive to managing chronic illness. It became readily apparent that I had to make drastic changes—and quickly. A solo short trip to San Diego was my first in many steps towards health. Not being a spiritual person or a meditative person or even a diet or exercise person, I had no framework with which to begin. I was going to the great unknown without a clue. First, I tried to get very quiet and make an attempt at prayer. Since this was quite new to me, I was not really sure how to go about it. But I tried my best at a desperate plea to make myself better and then I sat by a stream and waited for an answer. A hummingbird came to visit while I was waiting on this cold

February day. I did not get an answer. Or, maybe I did get one and did not know how to interpret it.

What I did get, however, was mad. I was mad that there was no one to help me, mad that I had to do this myself. Just like always, I thought. And with that thought, I was off and running. The issue was that I felt out of control. I asked myself, "So what do I have control of?" Immediately getting up from the cold ground, I headed to a café for some warm tea. With my tea to sip, pen and paper to write, the list flew out of me and onto the page.

What could I control? I made a list: diet, exercise, meditation, acupuncture, alternative therapies, and read books about the disease process. I had deconstructed my entire lifestyle that weekend and started building a "new and improved" Heather. I came up with my 20 tools for managing chronic illness. With these tools in hand and my good health demonstrating effectiveness in my life, I started speaking at support groups, writing articles and motivating others. Yet, my search for knowledge was merely just whetted.

I started to take classes at the American Institute of Holistic Theology, AIHT. They offered courses in many varied healing modalities. Class after class, I was absorbing more knowledge. Using myself as a "guinea pig," I tried different techniques and evaluated their effectiveness. I experimented with color therapy, sound therapy, and various spiritual disciplines, from pendulum therapy to astral projection. Some, as to be expected, were more effective than others. Some courses did not resonate with me, but I would approach them all with an open mind because, as I would soon discover, I never know what is in store.

In the years taking courses at AIHT I have earned my BS, MS, and PhD in parapsychic science. During the master's program, there was one particular course that would have a significant impact on my life—a course on hypnotherapy. In this course, there was a relatively small portion dedicated to past-life regression therapy. This section piqued my curiosity. Was there really something to this type of therapy? I had never given the idea of reincarnation much thought. I figured that I had too many issues in this life to worry about another one. And to be completely truthful my "woo-woo" alarm would turn on when hearing others talk about reincarnation, channeling, astral travel, etc.

The course on hypnosis related some astounding case studies with almost miraculous healings. Clients were sometimes healed in one session after years of suffering, whether it was from emotional or physical issues. They reported spiritual awakenings, they changed careers, and they forgave difficult and strained relationships. These clients reported that their lives changed significantly for the better. Yes, the sessions were usually quite intense with tragic and often violent death scenes, but after "re-living" these events, the clients left with a new perspective and deep peace.

So, out of sheer curiosity and a healthy dose of skepticism, I decided to find out for myself what this methodology was about. After some research, and referrals from other hypnotists into the qualified practitioners in my local area, I chose Dr. Donna Kannard. Donna Kannard has a PhD in clinical hypnotherapy and specializes in past-life regression. What happened next, during the first session, dramatically altered my life course.

When I was young, I did not like anything touching my neck. I would not wear turtlenecks, scarves, or chokers. Interestingly, in 2000, a tumor was found on my throat area that required surgical removal. The panic I went through, as the surgeon's blade lowered toward my throat, was more intense than would have been normal. It did not conform to my predisposition for "sucking it up." Today I wear a scar on my neck from that event.

I have also had recurrent issues with losing my voice, and many days, for no apparent reason I had absolutely no voice at all. This is despite three medical evaluations and speech therapy. Furthermore, a few years ago I experienced an untoward reaction to a narcotic medication that sent me into a panic, the likes of which I had never known. I became paranoid and thought my throat was closing up. I was convinced I was dying. I was fine, but emotionally it took me days to shake that unnerving fear.

I went to see Dr. Kannard on a Thursday afternoon. On that day, just like many others, I had no voice. I could only whisper. Dr. Kannard verbally coached me to relax my muscles, using guided imagery to take me to a deep level of hypnosis. She guided me to relax my muscles starting at my head and ending with my toes. Next, she suggested that I imagine walking down a beautiful staircase . . . down . . . down . . . down. With each step down I would feel more and more relaxed, and I did. At the bottom of the stairs I would find myself in a lovely garden, exactly to my liking. I could see the garden in full detail in my mind's eye. Hypnosis seemed to come easily for me. Finally, when completely relaxed and resting in the garden, she had me visualize a library.

The idea of a library was comforting and familiar to me, as I have always loved books and reading. Dr. Kannard suggested that in one of the rows of books there was a section with my name on it. I noticed that there was an area of books in which "Heather" was printed on the bindings. One particular book seemed to beckon me. I pulled this book out and sat down at a desk in this imaginary library. As I opened the yellow, frayed pages, one page caught my attention. On the right side of the open page was a picture of a knight standing in front of a white horse. The horse was kneeling low. The knight was in chainmail, holding a shield in his left hand and a sword in his right. I could see the symbol or heraldry clearly that was displayed on his shield. His hair was wavy, dark, and partially covering his right eye. His face was drawn, thin, and his color was ashen.

Dr. Kannard instructed me to go into the picture, and within an instant I became the knight. I was no longer Heather. I was suddenly feeling the extreme weariness of battle and lack of food, yet I also felt a deep sense of duty and honor to a cause. I had a sworn duty to fight, but also an uneasy feeling that it was not a just cause. I was unafraid to die and would do so with honor. I had no sense of a woman in my life, only of my love and tenderness for my horse.

Dr. Kannard asked me to go further ahead in time. The next scene I saw was me in combat. The ground was green, and I was on foot, fighting with my sword tight in my right hand. My thought, as Heather, was "shouldn't I be on my horse?" But no, I was on foot. It is such a strange experience to have two thoughts in my head at the same time. Lying in the therapist's chair, my right hand was clenched tight and held up as if holding the sword, and my left was holding my imaginary shield. I was unable to unclench my fists.

Suddenly I was choking on blood. I'd been stabbed in the throat. Lying on the chair in the office, I felt as if my throat was closing up, I was gagging and coughing. Dr. Kannard instructed me to pull out of the body and watch the scene as if on a movie screen. I tried to do as she suggested. The sensations felt so real, the pain, the choking; I couldn't remove myself from him. So, I continued to cough and gag until finally the pain stopped. I felt as if I were floating. Peace. Peace is all I knew. My hands relaxed as I released my grip on my imaginary weapons. This peace I have never known. No pain, no struggle, just an all-encompassing serenity combined with a lovely floating sensation. There was no tension in my body. I felt freer, less constricted and eternal. I had a sense of connectedness with all life and people. I could stay there forever. But Dr. Kannard was talking to me. I didn't want to let go of my knight. I was still feeling him and didn't want to leave, but she guided me out of that lifetime and into the present.

The session ends and I am overwhelmed. I hug her. As I'm thanking her, we both notice that my voice is clear and strong! I started out hoarse and at a whisper. Now I have a hearty voice again. I am ecstatic. Could it be that reliving the knight's death has resolved my vocal problems?

In the days following, I awaken with a clear voice. Every time I answer the phone, I am pleasantly surprised. I become consumed with a desire to learn all about knights and research heraldry. I finally understand why I like certain cultures, music, art, and symbols and why I behave in particular ways and have specific quirks, ethics and a code that predate me. I have a better understanding of my fears and my strengths and where they originated. And although the healing benefits are surprising and I have a greater appreciation for my specific traits, this is not the revelation that I am still in awe over.

After these experiences I read many books about knights, heraldry, and medieval history. Today I am still curious, but less so. To focus on the minutia and details is missing the bigger picture. To me, the real meaning, the real message is more profound than just the knowledge that I was a knight or any other person. If we accept the idea that we are more than just what this current life tells us we are, the implications are far reaching.

One implication is that I am immortal. I will experience many other lives with many of the same souls I have been with before. Furthermore, I aware that I am not the person I know as "Heather." It is as if I am wearing a "Heather" suit and next time I may be wearing a "Henry" suit. This helped me to step back and look at conflicts and crises more objectively. I am just an awareness experiencing Heather's life.

These changes of perception and perspective also affected the way I look at disease, chronic illness, and crises. It has affected the way I look at Heather's body, the way I look at incidents and Heather's reactions to them. "My" chronic illness is no longer personal to me. I no longer "own" it. It is merely something that Heather is experiencing and there is likely a reason for her to experience it. It is a lesson that must be learned from living with disease. This helps me to remove myself from the personal impact and to detach from the emotionality of it.

Contemplating what I felt were truths for me and integrating my new belief system into my being made me feel more connected, more peaceful, and changed my perspective on life and death. Of course, these are my truths and not necessarily yours, but I give you this story to illustrate how I forged a new path for myself.

### Heather's Bio:

Dr. Heather Friedman Rivera obtained her PhD in parapsychic science in 2011, her doctorate in law in 2002, and is a certified clinical hypnotherapist specializing in past-life regression. Heather is also a registered nurse with over thirty years of experience. She served on the board of directors for the International Association for Regression Research and Therapies (IARRT) as co-director of Research. She is a reviewer/editor for the *International Journal of Regression Therapy*, and co-founder of an organization for advancing past-life research (PLR Institute). Heather has written nine books, including the Amazon best seller, *Healing the Present from the Past: The Personal Journey of a Past Life Researcher.* She currently focuses on writing fiction/young readers, in addition to book coaching, mentoring, editing, and facilitating writing workshops.

Heather may be contacted at (808) 344-9894 or https://www.heatherrivera.com/.

### Scientific Overview

### Bob Davis's Story:

For some unknown reason, the "unexplained" has been an

area of interest since childhood. I don't know why my analytic-like perspective dominates the way I interact with the world. Maybe it was cultivated when I was a young child. Besides playing sports and looking over my suspicious shoulder for possible dangers while walking the streets in the South Bronx, I often played chess, and read almanacs, baseball statistics, and books on unidentified flying objects (UFO). Maybe these areas of interest served as a subconscious way to escape the torment of an intensely angry personality in the form of an older brother who sought delight by instilling daily fear in family members. Whatever the driving force may have been, I was a "closet ufologist" while pursuing a doctorate degree in sensory neuroscience at the Ohio State University and then as a professor at the State University of New York for over three decades. Writing articles, receiving major research grants, and presenting papers at national and international conferences consumed much of my life while I was married and raising two beautiful children.

When my wife and I observed the sudden appearance and then disappearance of two orange orbs in the night sky in Sedona, Arizona, in 2012, my interest in UFOs skyrocketed to the point where I felt compelled to write my first book: *The UFO Phenomenon: Should I Believe?* Fortunately, my retirement in 2014 afforded me ample time to continue using my left-brained approach to life to focus on research and writing about an area of considerable interest—the unexplained. And so, I lectured on UFOs, was interviewed on numerous shows, and even co-authored a paper on UFOs published in the *Journal of Scientific Exploration* in 2018. I'm still a researcher, but different in some respects.

My obsession with the "unexplained" led to my second book: *Life after Death: An Analysis of the Evidence*, published in November 2017. This book was inspired, in part, by an unexplainable and mind-shattering experience while in Byron Bay, Australia, in January of 2017, to lecture at two conferences. I was introduced to a woman who claimed to access and channel "off-world" energies to create a spiritual awakening. My longstanding curiosity in the "unexplained" took hold but never did I realize my life would change from that moment forward.

While I was relaxed with my eyes closed, she began speaking about energy, timelines, DNA, time and space, the greater good, cleaning and healing different parts of the body, negative and positive entities, past lives, and emotional traumas, that seemed to perplex, yet maintain my intense interest. Suddenly, an invisible energy seemed to take control which facilitated both physical and psychological effects. My body began to respond involuntarily to her words and intentions. I coughed excessively as my facial muscles twitched rapidly and my head turned repeatedly from side to side. Moments of exhaustion mixed with feelings of high energy, euphoria, confusion, transcendence, and awe over-whelmed my world. As my head turned violently, I thought of Regan, the child in the movie *The Exorcist*, and wondered if I, too, was possessed by an entity.

Both confused and amazed by my extreme, uncontrollable, and involuntary behaviors, my analytically-based mind sought answers to what seemed to be an extraordinary, personal experience that defied all logic. And in the following few weeks, my perspectives on life and reality changed. Blissful feelings combined with the anxiety of a spiritual crisis became an emergency in need of control. It seemed like a non-ordinary state of consciousness which I questioned as being either a very positive transformative outcome, a delusional disorder, or both. Except for the confusion and associated anxiety, I never felt better. It was like seeing the world through the eyes as a child again; I was amazed at the wonder of it all. A metamorphosis occurred in the form of a shift in self-identity—from an over-inflated, ego-driven researcher who strived for recognition from the scientific community into a less self-centered, sensitive, and ineffable, aware individual. I felt liberated from the complexity of my over-analytic and conditioned brain. I believed that "I" was now in control—more of me and not my brain. Being able to experience oneself as awareness itself, and experiencing greater unity with myself evoked a newfound realization of who and what I knew myself to be—a shift in perspective out of the ego as me and into something more true and unlimited, that Life and I are one.

This experience may be described as a form of self-transcendence, a spiritual awakening and crisis which needed time to integrate. I read books and spoke to those who I thought had keen insights about what happened to me. Answers didn't seem to help much. Words alone, while cathartic at some level, could not shake the confusion and anxiety triggered by my persistent questioning of "what happened to me." My wife was also scared and was afraid how this all might affect our marriage. But time alone helped integrate the experience which calmed the newfound "me."

I came to a pronounced realization that there is indeed "something" more to life that scientific principles cannot explain. Since Byron Bay, I wonder more than ever why life evolved on our pale blue dot in space. And maybe even more importantly, I question the purpose of why I am conscious or aware of being aware, and why am I even alive. How did conscious experience and the subjective sense of "I" arise from lifeless matter? What is the nature of my feelings of love, enjoyment of nature's beauty, and the ability to plan future events? Does our three-pound physical mass of brain tissue facilitate this all, or can it be just an aspect of nature not to be questioned and simply just "is"? And researching what governs and regulates that "something" should be considered an important endeavor by mainstream science. This is why I wrote the book *Life after Death: An Analysis of the Evidence*. After all, mainstream science contends that once brain activity stops, the hard drive crashes and the screen fades to black forever. But how do they know that? What evidence exists that says death is the grand finale? I needed to try to find my own answers and am driven to do so in my own unique way. I may not be successful, but this experience, for some reason, and in some way, has pointed me in this direction. The main thing is that I feel right doing it. And if this is all a delusion, then so be it, I'll take it any day over the old "me."

So now, with this reconditioning behind me, I am researching and writing another book with one question as a central theme—that is, do some see a different reality or do they see this reality differently? I ask this, in part, for self-serving reasons. After all, I used to turn around frequently while walking the streets of the South Bronx as a kid, which persisted in varying anxiety-driven ways, until now. Although my head frequently turned back as a child, and especially in Byron Bay, it is now faced forward with a new spirit and heart. And I hope those around me can appreciate that. At least, I do.

**Bob Davis Bio:**

Dr. Davis is an internationally recognized scientist in his field and served as a professor at the State University of New York for over 30 years. He graduated with a PhD in sensory neuroscience from The Ohio State University, published over 60 articles in scholarly journals, lectured at national and international scientific conferences, and was awarded several major research grants. He has also written two books, entitled *Life after Death: An Analysis of the Evidence*; and *The UFO Phenomenon: Should I Believe?* Dr. Davis has lectured on these topics at both national and international conferences.

His website is www.bobdavisspeaks.com.

# Chapter 9
## Change is in the Air

*Our greatest blessings come to us by way of madness,*
*provided the madness is given us by divine gift.*[1]
*-Socrates*

*Historically, society in general has regarded people who talk*
*to God as being holy. But if God talks to you, you're*
*considered insane.*[2]
*-Thomas Szasz, Psychiatrist*

Medical schools continue to educate students in the traditional model of brain-based consciousness. This approach graduates physicians, nurses, and psychiatrists ill-equipped to offer empathetic or practical support to those who have experienced extraordinary phenomena. Conventional medicine maintains that NDEers, OBEers, UAP-related contactees, and those who recall past lives are either delusional, experiencing dissociative episodes, or reconstructing imagined events. Thus, the medical field is either unwilling or unable to validate the reality of atypical phenomena. As a result, experiencers feel isolated, misunderstood, or plain "crazy." They struggle to understand and integrate their experiences in a non-supportive and dismissive environment. Although the authors previously discussed the necessity for compassionate psychiatric care in integrating spiritual crises, a more detailed discussion is necessitated.

Erica McKenzie has over twenty-five years of nursing experience. Following her NDE, she identified the need for enhanced education regarding such extraordinary experiences and implementation of a spiritual component to healing protocol.

### Erica's Nursing Rotation

I can still recall the hospital's rundown psychiatric floor while doing my nursing psychiatric rotation. There were no state-of-the-art treatment rooms or new equipment to be found. Instead, the air felt cold and heavy, and reeked of mildew. The patients looked lost, disheveled, unkempt. and sadly forgotten. I was an observer, researcher, and student. As I conducted my assessments, I noticed the relationship between the staff and patients was distant, superficial, and somewhat edgy. I couldn't help but sense this existed due to the clinician's perception that their patients were "abnormal." Therefore, it was no surprise when our clinical instructor sat us down at the end of the shift and stated, "It's important to appear observant and to carry out our duties as if you were listening. Mental health care provides people the opportunity to lead a normal life like everyone else, and these people are not normal," she declared.

She also said not to believe what the patients were sharing because they were not speaking truth, but simply attempting to manipulate staff members. I knew this was wrong, although at the young age of eighteen, I didn't have the wisdom to know how I knew. I just couldn't wrap my head around the fact that it was easy for numerous staff members to be so heartless. I was told patients were crazy. Yet, several of them appeared to have the capacity to communicate with something others couldn't see. I felt that many of them were medicated because of it. It was unacceptable to me.

As the group discussion unfolded, I began to understand how this preconceived notion had come about. Our instructor was presenting the cases of several patients who had been involuntarily hospitalized. Applicable therapy, including psychotherapy, occupational therapy, tranquilizers and antidepressants, had been tried, yet none seemed to produce a therapeutic effect. Some of these patients were being considered for transfer to the long-term ward, which meant spending countless more days, if not years, amidst chronic psychotic patients.

As the weeks passed, I was able to listen to many of the patients' issues, which were all too often complicated with a challenging combination of symptoms I had never encountered before. All of the patients had such unique circumstances, yet I uncovered a common theme. Each person at some point in their life had become a victim of an unfortunate circumstance. In numerous patients, this created more disruption and increased chaos in their lives, resulting in choosing coping mechanisms such as alcoholism, drug addiction, or a string of unhealthy relationships. This in turn caused intense side effects, such as depression, suicidal tendencies, violent outbursts, psychological conflict, guilt, and erratic impulses.

By the end of my psychiatric rotation, I had become completely perplexed. Why was there no mention of non-medicated psychotherapy or emotional support for the individual going through a "spiritual emergency" or crisis? I learned in medical school that medication was the most "therapeutic" way to treat "crazy behavior and delusions." Yet, when conventional interventions failed, and patients faced a lifetime in a psychiatric ward, where was the patient advocate? The advocate who would take a stand and say, "Since this isn't working, perhaps we should actively participate in holding a safe place for these individuals to be heard and supported instead of pathologized and labeled as crazy. What if, in doing so, this act of kindness would become the needed 'detox' and then a positive therapeutic protocol could be implemented."

I grew up believing that the purpose of medicine was to heal the sick. However, years later, I found myself in the exact same situation as some of the above-mentioned patients— medicated and pathologized because I had an experience the medical professionals couldn't explain. Call my near-death

241

experience crazy if you must—you won't hear me challenge it. But what does being crazy really mean? And does it constitute a solution if the cure comes in the form of a pill with a diagnosis of psychosis attached to it? Perhaps I was just damaged and lost or just maybe I was having a "spiritual emergency."

After my NDE, I realized that these patients needed to be understood by an educated, empathetic, and nurturing staff. I couldn't help but feel that the greater number of drugs administered to the patients acted as a Band-Aid, diminishing or concealing their symptoms temporarily. I sensed the drugs affected the ability of many of the patients to think clearly. Even more alarming, I sensed the drugs were *inhibiting* the ability of numerous patients to communicate with the Divine.

Trying to integrate my NDE was my "crash" course in mental health, beginning with my "crazy behavior and delusions." I realized I was given a multidimensional lesson. Everything that happened to me up until the day I died, including my trip to Heaven, Hell, and the psych ward, was preparing me to help both myself and others. That help would encompass educating the medical community about near-death experiences, the presence of God, and our ability to connect with our Creator. Additionally, an acute spiritual crisis, if provided with the right tools, support, and environment, could ultimately be profoundly healing.

During my NDE, I learned that we are spiritual beings having a human experience. I can't conceptualize the countless individuals who have experienced spiritual events and have been medicated and dismissed because of it. When does communicating with the unseen or spiritual realm validate a diagnosis of psychosis? I feel society, and especially the medical community, is in critical need of NDE and other non-ordinary phenomenon education from a spiritual perspective. This would increase its receptiveness to and support of experiencers, and help assist in differentiating the psychic

242

from the psychotic.

Obviously, there are individuals who experience hallucinations and require psychological treatment and pharmacological therapies in order to navigate life successfully. However, NDEs and non-ordinary phenomena which simulate acute "spiritual crises" should not necessarily be grouped with those who suffer true hallucinations. I strongly believe the medical system, as a whole, is over prescribing medications. I realized that God sent me back to Earth with the mission of contributing to improving the human condition. After all, when all humans take their last breath and die, they leave their physical bodies behind and only their souls continue the journey. Therefore, perhaps it's worth taking a more serious look at the value of the human soul and its Earthly contribution.

I'm not passing judgment on any institution or medical professional. That's not why I was sent back. Instead, I am identifying, through my experiences and lessons learned, that there is an urgent need for education on this subject. Because of this lack of body, mind, and spirit education, talking about my trip to Heaven, Hell, and God caused me to be checked into a psych ward against my will and be forcibly drugged. Those drugs began to change me, not heal me, because they were not the appropriate medical treatment modality. Yet, what I experienced was real. It happened. My NDE, "spiritual crisis," and aftereffects were a vital component to my healing—it was the exact medicine I needed.

Years of research and personal experience of anomalous phenomena have convinced me that current beliefs of medical professionals concerning the human psyche and con-sciousness require much needed revision. Many of the conditions that are diagnosed as psychotic and blindly treated by suppressive medication are, in my opinion, actually difficult stages of profound self-transformation and of spiritual awakening. If they are accurately understood and supported in a positive environment, these psycho-spiritual crises can result in emotional and psychosomatic healing, miraculous psycho-logical changes, and consciousness evolution. Calling such experiences an emergency or crisis defines the less than desirable nature of these states, yet at the same time, implies their positive healing potential. Such experiences are usually followed by an experience of psycho-spiritual awakening.

While working as a hospice nurse, I had numerous patients report a connection with the Divine. In fact, one patient in particular felt she was in direct connection with her dead husband, as well as with a relative who had died unexpectedly two years before. She said she longed to be reunited with them. She shared her visions of dying people and felt at times she too was dying. After everything I had experienced on a personal and professional level, I didn't hesitate to share that I felt it was possible to experience death figuratively without actually dying physically. I was able to offer her the compassionate support she needed, yet which is so lacking in the current medical curriculum.

When speaking at the International European Conference held in Hungary, I had the privilege of sharing the stage with afterlife expert Stanislav Grof, MD, a psychiatrist with over sixty years of experience in research of non-ordinary states of consciousness. Grof believes encounters with the paranormal are all too common in people who have experiences of altered consciousness or are undergoing what he proposes is a "spiritual emergency." He maintains that experiencers may be diagnosed as "psychotic." However, their behavior is not necessarily symptomatic of disease in the medical sense. "We view them as crises of the evolution of consciousness, or 'spiritual emergencies'; comparable to the states described by the various mystical traditions of the world," stated Grof.

He further suggested that, "A large group of spontaneous episodes in non-ordinary states of consciousness, currently diagnosed as manifestations of serious mental diseases and treated by suppressive psycho-pharmacological medication, are actually difficult stages of a process of psycho-spiritual transformation. If they are properly understood and supported, they can produce positive therapeutic, transformation and even revolutionary potential."

As I was listening to Dr. Grof's words, my jaw suddenly dropped and my heart began pounding out of my chest. Here I was, a medical professional and expert representing the NDE community at a global convention. Yet, in that moment I felt faint as I recalled my own NDE. I began to relive the first moments following it in hyper speed:

The emergency room physician had made the decision to admit me to the hospital for overnight observation. I tried to speak but nothing came out; I was too exhausted to even try. I only remember drifting in and out of consciousness. The next time I woke, I had never felt so alone. I didn't know how much time had passed. I didn't feel good and I was desperately trying to make sense of my NDE. At that point, a doctor walked into my hospital room.

"Mrs. McKenzie, how are you this morning? I'm Dr. Imanass."

My stomach cramped. Oh God, here it comes! Like a gushing unexpected wave, I was going to throw up. I reached for the pink bedpan to catch my vomit. But then I realized it wasn't vomit coming up. It was my voice. I was throwing up my voice. "Doctor, doctor, oh my God! I have to tell you where I just came from!" I said. The words just flew out of my mouth and I had no control over what came out of me. "I've just been to Heaven, Hell, talked with God, and He showed me so many things." I continued.

The doctor cut me off immediately, turned, and hurried out of the room. He didn't acknowledge one word I said. He didn't even complete my needed health assessment. I'd just been to Heaven, Hell, and back, and the first person I told couldn't run away fast enough. A doctor is supposed to help you. What had I done wrong?

The door to my room was still ajar, and a nurse who had been standing in the hallway charting her patients peeked inside. She entered the room and sat down on the edge of my bed. As she leaned in toward me, she pressed her finger to her lips and told me to be quiet and listen. She said that she had worked for that doctor for years and that he was an atheist. She believed that sharing my NDE may cause him "to cross a professional line" and determine my "mental status" from a personal, not professional level.

She told me that it wasn't the right time to tell my story. She said she could get fired for even discussing such matters with me yet stressed that "It's important to share your story." I knew she was afraid for her job and she was helping me at great personal risk. I then realized I was in the psych ward.

In the distance, once again I heard Dr. Grof's voice pulling me back to reality. "In view of the absence of a clear consensus regarding the causes of functional psychoses, it would be more appropriate and honest to acknowledge our complete ignorance as to their nature and origin and use the term disease only for those conditions for which we can find a specific physical basis. Thus, we can open the door to novel approaches to at least some functional psychoses, yielding new perspectives that differ from the medical view of disease. This includes first considering treatment with a non-pharmaceutical approach."

I remember thinking, "Where was Dr. Imanass when I was checked into the psychiatric ward against my will, for trying to share my NDE? What would the outcome of my experience as a patient have been like if my attending physician held the same beliefs as Dr. Grof?"

### Psychiatry Almost Drove Me Crazy

By Paul Levy, pioneer in the field of spiritual emergence, and author of *The Quantum Revelation: A Radical Synthesis of Science and Spirituality* (2018).

I am a survivor of severe psychiatric abuse. There was a year or so in the early 1980s when I was in and out of psychiatric hospitals at least four times. During my visits to the hospital I was in the midst of a spiritual awakening that I was struggling to contain that was triggered and complicated by extreme psychological abuse at the hands of my father, who was a very sick man. I was suffering so deeply from the psychic violence perpetrated upon my mother and me by my father that it was making me "sick." One of the most difficult parts of my ordeal in the hospitals was not being listened to by the psychiatrists, either about the abuse by my father or the spiritual awakening. Spiritual emergences/emergencies oftentimes become activated because of a deep experience of wounding, abuse, or trauma.

In its initial stage, a spiritual awakening can look like and mimic a nervous breakdown, as our habitual structures of holding ourselves together fall apart and break down so that a deeper and more coherent expression of our intrinsic wholeness can emerge. The spiritual awakening aspect of my experience was so off psychiatry's map that it wasn't even remotely recognized. Instead of hearing me, about either the abuse or the awakening, I was immediately pathologized and labeled as the sick one. Being cast in the role of the "identified patient," I was assured that I was going to be mentally ill for the rest of my days, as if I was being given a life sentence with no possibility for parole, with no time off for good behavior. The fact that I wanted to dialogue about this and question their diagnosis was proof, to the psychiatrists in charge of me, of my illness. The whole thing was totally nuts.

Fully licensed and certified by the state, the psychiatric

system's abuse of its position of power was truly unconscionable. What the profession of psychiatry was unconsciously enacting was truly crazy-making for those under their dominion. I was lucky to escape the psychiatric world with my sanity intact. Many others are not so fortunate. In not listening to what I was saying about the abuse being perpetrated by my father, and pathologizing me instead, psychiatry was unwittingly protecting my father. It was as if the field of psychiatry had become subsumed into unknowingly becoming an instrument for a deeper, archetypal process of "protecting the abuser" to play itself out in form and in real time.

Having my father be caretaken by those in a position of potential authority over him, combined with my being solidified as being sick by those in authority over me was a doubly sickening experience. It was only years later, long after I had left the psychiatric community, that it began to come out in my family that my father was both criminally and morally insane, a genuine psycho-sociopath who was truly a danger to others. (Note, in trying to decide whether to use the word "psychopath" or "sociopath," I've decided to create a new term, "psycho-sociopath," to describe my father's condition. It "sounds" right.). In not listening to me or recognizing the reality of my father's virulent pathology, the psychiatrists were complicit in the abuse.

A dreamlike image comes to mind that actually happened to me that speaks louder than words. I'm locked up in a mental hospital in the midst of having a full-blown spiritual awakening. I'm sitting in my room in the hospital meditating. I am, moment by moment, watching my thoughts arise and in their very arising naturally dissolve back into the spacious emptiness from which they arise. I am dis-identifying from my thoughts, and more and more recognizing that I can just rest in the spacious emptiness which is our true nature. In addition to freeing my consciousness from the limitations of the conceptual mind, meditation is the one thing I've found which is healing the abuse from my father. Into the room comes the psychiatric ward attendant, who surreally happens to be one of the big high school basketball stars in the city in which I grew up, and he is stopping me from meditating. I am not allowed to meditate. He has it in his notes that meditation caused my illness.[3]

Satwant K. Pasricha is the former head of the Department of Clinical Psychology, National Institute of Mental Health and Neurosciences, Bangalore, India. Since 1973, Pasricha has investigated over five hundred cases of children who recall past lives and has extensively researched near-death experiences. She considers it imperative that clinicians gain a deeper and broader understanding of extraordinary experiences/ers. As she explains:

> There may be phenomenological similarities between parapsychological experiences and psychiatric conditions. However, with adequate knowledge and training a detailed evaluation would show that the two conditions are entirely different and require different management strategies. A psychiatrist who is not at least open-minded about the possibility of paranormal experiences will almost certainly be unable to distinguish psychopathological from para-normal and equally unable to assist the occasional person who is perplexed about unusual experiences that he would like to report and discuss with someone outside his family.
>
> In sum, para-psychology is relevant to psychiatry in the following spheres: First, its knowledge would assist mental health professionals in differentiating para-normal

experiences from psychopathological phenomena leading to adequate treatment strategies.
Second, it would enhance understanding of
certain medical, psychological and psychiatric
disorders that cannot be explained in terms of
currently available theories of the genetic or
environmental influences. Third, it would
facilitate advancement of knowledge in
brain/mind relationship.[4]

Further training for medical professionals is urgently
needed. In order to create caregivers who are compassionate
and open-minded towards anomalous experiencers, medical
curriculum must consider the inclusion of coursework in
spirituality, consciousness studies, and open-minded, compassionate *listening* skills. There are many mysteries in life
which are inexplicable to, and incompatible with, traditional
scientific and medical paradigms. Yet, because something is
unfathomable to conventional minds does not invalidate its
truth or very existence.

Thankfully, numerous universities have initiated training
programs and research studies designed to provide a broader,
more compassionate understanding of extraordinary
experiences/ers. Colleges in both the United States and abroad
are beginning to implement, or have recently completed,
advanced research programs which investigate phenomena
concerning the nature and inter-relationship of the mind,
brain, and consciousness. Certain initiatives have been met
with skepticism or, at times, ridicule. However, universities
continue to challenge traditional education by offering novel
curriculum and cutting-edge research. Several schools
currently offer or are affiliated with such programs. These
currently include but are not limited to The University of
Virginia's Division of Perceptual Studies (DOPS); The
University Of Arizona's Center for Consciousness Studies;
John Hopkins University School of Medicine; The University

of Edinburgh; Cornell University; Stanford University; Duke University; and the University of Southern California (USC).

DOPS, originally known as the Division of Personality Studies, was founded in 1967 by the late reincarnation researcher Dr. Ian Stevenson. Its research explores such phenomena as reincarnation (most specifically through its focus on children who claim to remember past lives); near death experiences (NDEs); apparitions and after-death communications; altered states of consciousness; as well as many other psychic (psi) experiences.

The University of Arizona's Center for Consciousness Studies was initiated to incorporate the fields of philosophy; cognitive science; neuroscience; social/ physical science; medicine; and the arts/humanities. Its goal is to investigate human consciousness in a fully integrated manner.

The University of Arizona created the VERITAS Research Program in 2006 to explore the possibility that human consciousness (or personality or identity) survives physical death. The project operated out of the Laboratory for Advances in Consciousness and Health (LACH) at the University of Arizona's Department of Psychology. Following its completion in 2008, LACH director Dr. Gary E. Schwartz expanded the VERITAS study into a broader, more comprehensive spiritual communications project called the SOPHIA Research Program. This ongoing program investigates communication with discarnate energies, such as spirit guides, angels, and even with a higher power or divine being, with the focus on "healing and life-enhancement." According to the LACH website, "consciousness exists and can be investigated "scientifically," and "consciousness is in a dynamic state of change and evolution." Schwartz's research projects at LACH also include investigations into such fascinating areas as quantum holographic consciousness, the universal intelligence hypothesis, and animal consciousness.[5]

Roland Griffiths, PhD, is a professor in the Departments of

Psychiatry and Neurosciences at the Johns Hopkins University School of Medicine. His principal research focus in both clinical and preclinical laboratories has been on the behavioral and subjective effects of mood-altering drugs. His research has been largely supported by grants from the National Institute on Health and he is author of over 360 journal articles and book chapters. He has been a consultant to the National Institutes of Health, and to numerous pharmaceutical companies in the development of new psychotropic drugs. He is also currently a member of the Expert Advisory Panel on Drug Dependence for the World Health Organization. He has conducted extensive research with sedative-hypnotics, caffeine, and novel mood-altering drugs. In 1999 he initiated a research program at Johns Hopkins investigating the effects of the classic hallucinogen psilocybin that includes studies of psilocybin-occasioned mystical-type experiences in healthy volunteers. Their drug interaction studies and brain imaging studies (fMRI and PET) are examining pharmacological and neural mechanisms of action. The Hopkins laboratory has also conducted a recent series of internet survey studies characterizing the effects of hallucinogen-occasioned mystical experiences, and Erica McKenzie served as the ambassador for their anonymous, web-based, global survey study regarding near-death or other non-ordinary experiences.

The Koestler Parapsychology Unit was established at the University of Edinburgh (Edin), Scotland, in 1985. Unbeknownst to many Americans, Edin is an extremely prestigious institution. In 2019, it was ranked as the eighteenth top university in the world.[6] Koestler teaches and researches parapsychology, focusing on the areas of psi hypothesis/pseudo-psi, the paranormal, and anomalous phenomena.

Cornell University conducted an eight-year research study (2002–2010). Headed by social psychologist Daryl Bem, researchers examined psi/precognition in over one thousand subjects. His findings strongly suggested that humans do possess physic and precognitive abilities. The results of Bem's study, entitled "Feeling the Future: Experimental Evidence for Anomalous Retroactive Influences on Cognition and Affect," was published in the American Psychological Association's *Journal of Personality and Social Psychology.*[7]

Stanford (SRI International) is a world-renowned private research university, currently ranked number two in the U.S. after Harvard University.[8] In 1912 it became the first United States academic institution to research extra-sensory perception (ESP) and psychokinesis (PK). From 1972 to 1991, numerous studies were conducted on remote viewing and psi, which indicated these phenomena could be scientifically researched. SRI published their findings in several mainstream journals, attracting sponsorship by both NASA and the CIA. In 1991 research was transferred from Stanford to the Science Applications International Corporation (SAIC), in Reston, Virginia, to initiate the Stargate Project, a code name for a classified army unit to explore the use of psychic ability in military applications. The project was terminated and declassified in 1995, after it proved useless to army operations.

Stanford is once again exploring cutting-edge research. Established in 1983, its Center for the Explanation of Consciousness (CEC) is undertaking an interdisciplinary approach to examining diverse theories of consciousness. The schools of psychology, linguistics, philosophy, computer science, communication, education, and its symbolic systems program are conducting research collaboratively.

### Princeton Study's Findings Face Ridicule

In February 2007, the *New York Times* (NYT) published an article concerning the closing of Princeton's Engineering Anomalies Research Lab (PEAR). As NYT science reporter Benedict Carey explained:

For over almost three decades, a small laboratory at Princeton University managed to embarrass university administrators, outrage Nobel laureates, entice the support of philanthropists and make headlines around the world with its efforts to prove that thoughts can alter the course of events.

The laboratory conducted studies on extrasensory perception and telekinesis from its cramped quarters in the basement of the university's engineering building since 1979. "For 28 years, we've done what we wanted to do, and there's no reason to stay and generate more of the same data," said the laboratory's founder, Robert G. Jahn, 76, former dean of Princeton's engineering school and an emeritus professor. "If people don't believe us after all the results we've produced, then they never will."

Princeton made no official comment.

PEAR had been an anomaly from the start, a ghost in the machine room of physical science that was never acknowledged as substantial and yet never entirely banished. Its longevity illustrates the strength and limitations of scientific peer review, the process by which researchers appraise one another's work.

As sociologist Harriet Zuckerman, author of *Scientific Elite: Nobel Laureates in the United States* and senior vice president of the Andrew W. Mellon Foundation explains: "We know people have ideas beyond the mainstream, but if they want funds for research they have to go through peer review, and the system is going to be very skeptical of ideas that are inconsistent with what is already known."

Dr. Jahn, one of the world's foremost experts on jet propulsion, defied the system. He relied not on university or government money but on private donations — more than $10 million over the years, he estimated. The first and most generous donor was his friend James S. McDonnell, a founder of the McDonnell Douglas Corporation.

Those gifts paid for a small staff and a gallery of random-motion machines, including a pendulum with a lighted crystal at the end; a giant, wall-mounted pachinko-like machine with a cascade of bouncing balls; and a variety of electronic boxes with digital number displays.

In one of PEAR's standard experiments, the study participant would sit in front of an electronic box the size of a toaster oven, which flashed a random series of numbers just above and just below 100. Staff members instructed the person to simply "think high" or "think low" and watch the display. After thousands of repetitions — the equivalent of coin flips — the researchers looked for differences between the machine's output and random chance.

Analyzing data from such trials, the PEAR team concluded that people could alter the behavior of these machines very slightly, changing about 2 or 3 flips out of 10,000. If the human mind could alter the behavior of such a machine, Dr. Jahn argued, then thought could bring about changes in many other areas of life

— helping to heal disease, for instance, in oneself and others.

This kind of talk fascinated the public and attracted the curiosity of dozens of students, at Princeton and elsewhere. But it left most scientists cold. A physics Ph.D. and an electrical engineer joined Dr. Jahn's project, but none of the university's 700 or so professors did. Prominent research journals declined to accept papers from PEAR. One editor famously told Dr. Jahn that he would consider a paper "If you can telepathically communicate it to me."

Brenda Dunne, a developmental psychologist, managed the laboratory since its inception and co-authored many of its study papers. "We submitted our data for review to very good journals," Ms. Dunne said, "but no one would review it. We had been very open with our data. But how do you get peer review when you don't have peers?"

Several expert panels examined PEAR's methods over the years, looking for irregularities, but did not find sufficient reasons to interrupt the work. In the 1980s and 1990s, PEAR published more than 60 research reports, most appearing in the journal of the Society for Scientific Exploration, a group devoted to the study of topics outside the scientific mainstream. "It's time for a new era," he said, "for someone to figure out what the implications of our results are for human culture, for future study, and — if the findings are correct — what they say about our basic scientific attitude."[9]

The current, ongoing Global Consciousness Project (GCP)

originated from PEAR research. The objective of GCP is to investigate the impact that worldwide, group consciousness has on the physical world.

Of particular interest, several universities are introducing coursework on UAP-related phenomena. Duke University's Osher Institute for Lifelong Learning is offering a course in continuing studies entitled "Encounter, Mystery, Myth?" The professor (whose name, at this point, remains anonymous) is a well-known UFO researcher who previously taught a ufology class at UNC-Chapel Hill. As he explains: "It's not the first time I've taught a course on UFOs. In the spring semester of 1996, when I was a professor in UNC-Chapel Hill's Department of Religious Studies, I gave a seminar called 'Special Topics in Mysticism: Heavenly Ascensions and UFO Abductions.' One of my colleagues indulgently labeled it 'a funny course.' He didn't understand. Whatever UFOs are, they aren't funny."[10]

Meanwhile, at the University of Southern California, graduate students in computer science are investigating thousands of documented UFO sightings to determine, "Are we alone in the universe"?

Currently, no traditional universities offer degree programs in ufology, but there are some online options. The International Metaphysical University offers six courses in Ufology Studies, including Introduction to Ufology taught by Richard Dolan, a well-known expert who also has a history degree from Alfred University. Two other online universities— the Centre of Excellence in the U.K. and the IMHS Metaphysical Institute—offer full degree programs in ufology.

Award-winning writer Paul Seaborn remarks on the significance of ufology coursework by stating: "What DOES it mean? It provides a little more legitimacy to a subject begging for it."[11]

Thankfully, research and education are introducing a new perspective of death and continuation of consciousness in both the scientific and medical communities. Hopefully, this awareness will continue, both improving the quality of healthcare, and graduating medical professionals empathetic to experiencers of anomalous phenomena. As Sir John Eccles, neuro-physiologist and Nobel Prize winner expressed, "I maintain that the human mystery is incredibly demeaned by scientific reductionism, with its claim in promissory materialism to account eventually for all of the spiritual world in terms of patterns of neuronal activity. This belief must be classified as a superstition. We have to recognize that we are spiritual beings existing in a spiritual world as well as material beings with bodies and brains existing in a material world."[12]

# Chapter 10
## Final Thoughts

*I don't think you can locate the source of consciousness. I am quite sure it is not in the brain, not inside of the skull ... It actually, according to my experience, would lie beyond time and space, so it is not localizable. You actually come to the source of consciousness when you dissolve any categories that simply imply separation, individuality, time, space, and so on. You just experience it as a presence.[1]*
Stanislav Grof, Psychiatrist

*Even the physicists and scientists who proselytize the materialistic model have been forced to the edge of the precipice. They must now admit to knowing just a little bit about 4% of the material universe they know exists but must confess to being totally "in the dark" about the other 96%. And that doesn't even begin to address the even grander component that is home to the "Consciousness" that I believe to be the basis of it all.[2]*
Eben Alexander, Academic Neurosurgeon and NDEer

*Neither the mind nor the world is divided into compartments.[3]*
Bernard D'Espagnat, Theoretical Physicist, Philosopher

*Reality is woven from strange, "holistic" threads that aren't located precisely in space or time. Tug on a dangling loose end from this fabric of reality, and the whole cloth twitches, instantly, throughout all space and time.[4]*
Dean Radin, Researcher at the Institute of Noetic Sciences

As lifetime experiencers, the authors maintain that NDEs, OBEs, UAP-related contact, and past-life recall are non-local, interconnected experiences. They occur beyond space, time, and our physical senses. They are not uncommon yet are consistently under-reported due to fear of ridicule, professional "suicide," confusion, and/or embarrassment. Experiencers unshakably believe that extraordinary phenomena are realer-than-real, existing outside of our earthly time continuum. Yet, the majority of humanity continues to operate in a three-dimensional world bound by time, space, and belief in localized consciousness. Why is this? Is there an answer?

Larry Dossey, MD, offers one explanation: "We don't know who first discovered water, but we can be sure it wasn't a fish ... Continual exposure to something reduces our awareness of its presence. Over time, we become blind to the obvious. We swim in a sea of consciousness like a fish swims in water. And like a fish that has become oblivious to his aqueous environment, we have become dulled to the ubiquity of consciousness."[5]

Dharma Master Hwansan Sunim, contributing writer to the Huffington Post, offers another: "Science, by its own definition, is the investigation of one domain, namely the observable material world. As a scientist, then, saying that you are only going to investigate one domain of phenomena does not logically mean that no other such domain exists ... in a word, it's bad science to say that no observations, except the kinds of ones that I make, are possible. It's ok to be uncomfortable with the unfamiliar, but can't we be honest about it?"[6]

Cardiologist Pim van Lommel asserts: "Scientific research into extraordinary experiences highlights the limitations of our current medical and neurophysiological ideas about the various aspects of human consciousness or self ... and the brain on the other ... For decades, extensive research has investigated localized consciousness and memories inside the brain, so far without success."[7]

It is time to bring about a revolutionary change in science—one which will serve as a catalyst to overhaul the outmoded nuts and bolts approach to extraordinary experiences. This will challenge the very heart of materialism, which is notoriously slow to embrace new paradigms. When presented with a multitude of anomalous reports, science must focus on the information provided by experiencers. It is obliged to examine, "What is Consciousness?" and "What is the true nature of our reality?" The answers and data may rattle conventional research. However, pushing the limits of current under-standing is the only way to instigate new, cutting-edge hypotheses. As physicist Nikola Tesla so aptly stated: "The day science begins to study non-physical phenomena, it will make more progress in one decade than in all the previous centuries of its existence."[8]

> *If you wish to upset the law that all crows are black ... it is enough if you prove one single crow to be white.*[9]
>
> *William James,*
> *Philosopher and Psychologist*

Experiencers are eagerly awaiting the day science is *wholeheartedly* committed to investigating the inter-connectivity of experiential modalities *via* non-locality. Hopefully, it will arrive at the same conclusion as the authors:

- Consciousness is non-local and eternal.
- Extraordinary experiences are positively life-altering.
- Extraordinary experiences are "pieces of the same non-local puzzle" and share numerous commonalities.
- The human energy body extends beyond the

physical body, time, and space—connecting with all else in a holographic universe.

- The *experiential* component of phenomena holds as much significance as the science behind it.
- Experiencers exhibit a unique personality prone to anomalous occurrences, which is non-pathological in nature.
- It is crucial to educate medical professionals in the following areas: spirituality/spiritual crises, death and dying, and compassionate psychiatric care.
- To provide a respectful environment which encourages experiencers to safely "come out of the closet."
- To provide improved and expanded support to experiencers in the following areas: medical/psychological care, support groups for both experiencers and their families, specialized therapy, forums, and conferences.

*"The important thing in science is not so much to obtain new facts as to discover new ways of thinking about them."*[10]
*Sir William Lawrence Bragg, Physicist*

I have experienced countless anomalous phenomena throughout my life, including spontaneous past-life memories, astral projection, precognition, and mediumship abilities. I have seen ghosts, orbs, experienced inexplicable electrical phenomena and UAP-related contact, received downloads of previously unknown information, and have observed objects dematerialize. I am a highly sensitive person who has experienced phenomena via a timeless, non-earthly dimension. Thus, I am absolutely convinced that we live in a multiverse in which everything is interconnected via non-locality.

Barbara

I have had anomalous experiences since childhood. When I was a little girl, I experienced communication from my maternal grandmother who had passed away. That experience was one my mother and I shared. I had OBEs from a very young age and didn't understand what they were. I thought everyone was having these experiences and wasn't even aware they were out of the ordinary until I was a teenager. In adulthood, I've had multiple NDEs and have had to remind myself to "stay in my body." It seems once you've been "out" it becomes second nature to transition between the dimensions. My near-death experiences definitively convinced me that consciousness is not held in the brain, like some sort of component, but rather, is accessible via the brain. My work as a nurse further solidified my belief that consciousness is indeed non-local. As patients neared death, it was clear to me they were communicating with and accessing other realms.

Penny

The night skies have always called to me. It has become an impulse to look up— not to look for UAPs, because I have seen so much more— but to feel the connection to the universe as it sees its own magnificence through my eyes. I have had a lifetime of anomalous experiences: of beings, of other worlds, a connection, and communication that reveals itself in purity and total transparency. Within us are the stars, and they are ours. I am me, in this here and now, but I am also there in what is unseen. We are multidimensional beings, a consciousness of non-locality, our minds are beyond our physical bodies, enabling us to experience other realms. The universe creates life—that is its purpose. Whether this life is in our own "denser" reality or in realms we cannot see, it is there, and it is real. It is within us to explore it, as it reaches out for us. All we must do is reach out and grab it.

Lynn

There really is so much to support the non-local view of consciousness—including, for nearly all of us, our own experiences.

# Resources

## General

1. *After-Death Communication*  www.adcrf.org
2. *American Center for the Integration of Spiritually Transformative Experiences*  https://aciste.org
3. *Alternate Perceptions Magazine*  http://www.apmagazine.info/
4. *Stanislav Grof, MD. Psychiatrist who researches non-ordinary states of consciousness.*  http://www.stanislavgrof.com/
5. *Institute of Noetic Sciences. Researches the intersection of science and profound human experiences.*  https://noetic.org/
6. *Stuart Hameroff, MD. Anesthesiologist and Director of Consciousness Studies, University of Arizona*  https://www.quantumconsciousness.org/
7. *Thomas Campbell website. Physicist, author, consciousness researcher.*  https://www.my-big-toe.com/
8. *International Academy of Consciousness. Organization devoted to educating and researching consciousness and human potential.*  https://www.iacworld.org/
9. *Institute of Applied Consciousness.*  https://www.conscioustech.com/

## Information/Personality-Prone Type Testing

1. *MBTI online quiz.*  http://www.humanmetrics.com/cgi-win/jtypes2.asp
2. *Information on the Gregorc mind-styles model*https://hollyclark.org/wp-content/uploads/2017/08/Anthony-Gregorc-4-Mind-Styles.pdf
3. *Gregorc online mind-styles quiz*

http://wp.auburn.edu/biggio/wp-content/uploads/2012/07/learning-style-test.pdf

4. Suggested reading: *The Hiss of the ASP: Understanding the Anomalously Sensitive Person.* David Ritchey, Headline Books, Terra Alta, WV, 2003. ISBN 0929915291

## Associations/Websites
## Near-Death Experience

1. *Near-Death Experience Research Foundation. Largest collection of NDEs in over 23 languages.* www.nderf.org
2. *International Association for Near-Death Studies. Annual Conference, information, research, and support groups.* www.iands.org
3. *P.M.H. Atwater. Internationally renowned near-death expert/researcher.* www.pmhatwater.hypermart.net
4. *Dr. Kenneth Ring. Professor Emeritus of Psychology at the University of CT and world-renowned NDE researcher.* www.kenring.org
5. *Life after Life Institute. Information concerning afterlife.* www.lifeafterlife.com
6. *Pim van Lommel, MD. Cardiologist and NDE researcher.* www.pimvanlommel.nl/en/
7. *Public Facebook group.* https://www.facebook.com/groups/returnfromdeath/
8. *Search for NDE experiencers groups near you!* https://www.meetup.com/topics/near-death-experience/
9. *Convergence of Science and Spirituality for Personal and Global Transformation. Founded by Eben Alexander, MD, and near-death experiencer himself.* http://www.eternea.org/

## OBEs

1. *Out-of-Body Experience Research Foundation.*

*Information, resources, experiences.* www.oberf.org

2. *The Monroe Institute. Nonprofit education and research organization, a preeminent leader in human consciousness exploration.*
   https://www.monroeinstitute.org/
3. http://www.astralinfo.org/
4. *Bob Peterson. Author, expert, out-of-body experiences.*
   *http://www.robertpeterson.org/*
5. *Jurgen Ziewe. Author, OBEer, and consciousness explorer. Website for those interested in out-of-body travel and exploring higher states of consciousness.*
   https://www.multidimensionalman.com/Multidimension
   al-Man/Astral_Travel_and_life_after_death.html
6. *Website of Robert Bruce.* https://astraldynamics.com/
7. *Website of Robert Waggoner. Past President of the International Association for the Study of Dreams (IASD) and co-editor of the online magazine The Lucid Dreaming Experience, the only ongoing publication devoted specifically to lucid dreaming.*
8. https://www.lucidadvice.com

## UAP-Related Contact

1. *Australian Close Encounter Researcher Network. Counseling, hypnotherapy, and research.*
   http://acern.com.au/
2. *Mutual UFO Network. World's oldest/largest UFO investigation and research organization.*
   www.mufon.com
3. *National UFO Reporting Center.* http://www.nuforc.org/
4. *National Government Archives, including Project Blue Book.*
5. https://www.archives.gov/research/military/air-force/ufos.html
6. *Beyond Presidential UFOs*

https://beyondpresidentialufo.com/

7. *SETI. Search for Extraterrestrial Intelligence. Searches for signs of transmission from civilizations on other planets.* https://www.seti.org/

8. *J. Allen Hyneck. Center for UFOs.* www.cufos.org

9. *Disclosure Project. Archived information disclosing facts on UFOs, extraterrestrial intelligence, and advanced energy and propulsion systems.* https://www.disclosureproject.org/

10. *Annual UFO conference.* https://ufocongress.com/

11. *Kathleen Marden. Leading UFO researcher, author, and international lecturer.* http://kathleen-marden.com/

12. *Science and technology research and investigation group.* https://thefieldreportscom.wordpress.com/

13. *Richard Dolan. One of world's leading researchers and writers on UFOs.* https://www.richarddolanpress.com/

## Past-Lives

1. *Brian Weiss, MD. Psychiatrist, world-renowned author, and past-life regression therapist* http://www.brianweiss.com/

2. *Brian Weiss, MD. Brian Weiss Institute.* http://www.brianweiss.com/contact-weiss-institute/

3. *Heather Friedman Rivera, PhD. Co-founder of the Past-Life Research Institute.* http://www.heatherrivera.com/plr-institute/

4. *Dr. Jim B. Tucker. Psychiatrist and Director of the University of Virginia Division of Perceptual Studies. Past-life research.* http://www.jimbtucker.com/

5. *Past-life Memories research.* https://psi-encyclopedia.spr.ac.uk/articles/past-life-memories-research

6. *The late Ian Stevenson, psychiatrist and ground-breaking past-life researcher at the University of Virginia School of*

*Medicine.* https://www.near-
death.com/reincarnation/research/ian-stevenson.html

7. *Find past-life meetup groups in your area.*
   https://www.meetup.com/topics/past-life/

8. *Carol Bowman. Renowned past-life regression therapist.*
   https://www.carolbowman.com/past-life-regression

9. *Carol Bowman Reincarnation Forum*
   http://reincarnationforum.com/

10. *The late Roger Woolger, PhD. Woolger specialized in past-
    life regression spirit release and shamanic healing. The
    Woolger Institute for Psyche and Spiritual Awareness.*
    http://rogerwoolger.org/

# Endnotes

**Chapter 1**

1 Michael J. Robey, "So You Want to Be a Channeler Too?" SelfGrowth.com. Accessed March 1, 2019. https://www.selfgrowth.com/articles/so-you-want-to-be-a-channeler-too.

2 Raymond Moody, *Life After Life: The Bestselling Original Investigation That Revealed Near-Death Experiences* (New York, New York: HarperCollins Publishers, 2001), 110 .

3 "What Is Matter? - Never Mind. What Is Mind? - No Matter." "Punch," London humor magazine. Accessed February 23, 2019. https://quoteinvestigator.com/2018/08/30/matter-mind/.

4 "Consciousness Science: A Brief Introduction." International Academy of Consciousness. Accessed June 10, 2018. https://www.iacworld.org/consciousness-science-introduction/.

5 Pim van Lommel, *Consciousness Beyond Life: The Science of the Near-Death Experience* (New York, New York: HarperCollins, 2010), xvii.

6 "Science and the Near-Death Experience." Near-Death Experiences and the Afterlife. Accessed March 4, 2019. https://www.near-death.com/science/research/science.html.

7 Van Lommel, *Consciousness Beyond Life: The Science of the Near-Death Experience*, ix

8 Stephen Michael Nanninga, "Transcendental Omnipresence." Of Quasars and Quanta. Accessed December 12, 2018. http://www.cosmiclight.com/ofquasars/omnipresence.html.

9 Mario Beauregard, Gary E. Schwartz, and Lisa Miller. "Manifesto for a Post-Materialist Science." 2014. Accessed

June 14, 2018. http://opensciences.org/about/manifesto-for-a-post-materialist-science.

10 Daniel Dennett, *Consciousness Explained*. (Boston, MA: Little Brown, 1991), n.p.

## Chapter 2

1 "Anomalous." Dictionary by Merriam-Webster, accessed December 10, 2018. https://www.merriam-webster.com/.

2 Kenneth Ring, and Christopher Rosing, "The Omega Project: An Empirical Study of the NDE-Prone Personality," *Journal of Near-Death Studies* 8, no. 4 (1990): 211-38.

3 Ibid.

4 "Dissociative Disorders." National Alliance on Mental Illness, accessed December 26, 2018. https://www.nami.org/learn-more/mental-health-conditions/dissociative-disorders.

5 David Ritchey, *The H.I.S.S. Of the A.S.P.: Understanding the Anomalously Sensitive Person* (Terra Alta, WV: Headline Books, Inc., 2003), 84.

6 James Lake, "The Evolution of a Predisposition for the Near-Death Experience: Implications for Non-Local Consciousness." *Journal of Nonlocality* 5, no. 1 (2017): 1-17.

7 Ritchey, *The H.I.S.S. of the A.S.P.: Understanding the Anomalously Sensitive Person*, 71.

8 Ibid., 69.

9 "Being Left-Handed." Left Handed Facts, accessed June 5, 2018. https://www.lefthandersday.com/tour/being-left-handed#.XGiiDaJKjIV.

10 Ritchey, *The H.I.S.S. of the A.S.P.: Understanding the Anomalously Sensitive Person*, 93.

11 "William Ralph Inge Quote." BrainyQuote, accessed December 10, 2018. https://www.brainyquote.com/quotes/william_inge_149271.

12 "Mind Styles." accessed November 13, 2018. http://web.cortland.edu/andersmd/learning/Gregorc.htm.

13 "What Is Your Dominant Thinking Style?" CanadaOne, 2014, accessed May 5, 2018.

https://www.canadaone.com/ezine/2013/what_are_your_d
ominant_thinking_styles.html.

14 "What Does It Mean to Be a Highly Sensitive Person?"
PsychCentral, 2018, accessed June 20,
2018. https://psychcentral.com/lib/what-does-it-mean-to-
be-a-highly-sensitive-person/.

15 Ritchey, *The H.I.S.S. of the A.S.P.: Understanding the
Anomalously Sensitive Person,* 173.

16 "Overwhelmed by the World?" Inner World, accessed
December 29, 2018.
http://www.aislingmagazine.com/aislingmagazine/articles/
TAM28/Overwhelmed.html.

17 Ritchey, *The H.I.S.S. of the A.S.P.: Understanding the
Anomalously Sensitive Person,* 44.

18 Ibid.

19 Susan Marie Powers, "Fantasy Proneness, Amnesia,
and the UFO Abduction Phenomenon," *Dissociation* IV, no. 1
(1991): 46-54.

20 Harvey J. Irwin, "Fantasy Proneness and Paranormal
Beliefs," *Psychological Reports* 66, no. 2 (1990): 655-58.

21 Nicholas Spanos, Patricia Cross, Kirby Dickson, and
Susan Dubreuil, "Close Encounters: An Examination of UFO
Experiences," *Journal of Abnormal Psychology* 102, no. 4
(1993): 624-32.

22 Joe Nickell, "A Study of Fantasy Proneness in the
Thirteen Cases of Alleged Encounters in John Mack's
Abduction," *Investigative Files* 20, no. 3. (May/June
1996). https://www.csicop.org/si/show/study_of_fantasy_pr
oneness_in_the_thirteen_cases_of_alleged_encounters_in_j.

23 Ritchey, *The H.I.S.S. of the A.S.P.: Understanding the
Anomalously Sensitive Person,* xi.

24 "The Physiology of Psychic People and Psychic
Ability," Core Spirit, accessed May 4,
2018. https://corespirit.com/the-physiology-psychic-people-
and-psychic-ability.

## Chapter 3

1 Jeffery Long, "Near Death Experience Overview," Near-Death Experience Research Foundation, 1999, accessed August 15, 2018. https://www.nderf.org/NDERF/Articles/NDE%20Overview. htm.

2 Charles Tart, "A Future for Dualism as an Empirical Science?" New Dualism Archive, 2006, accessed December 17, 2018. http://www.newdualism.org/papers/C.Tart/Tart-Dualism.pdf.

3 Ted Row, "Definition of Unidentified Aerial Phenomena, Uap." National Aviation Reporting Center on Anomalous Phenomena, accessed January 31, 2019. http://www.narcap.org/Blog

4 Ryan A. Zeigler, "Lunar Rocks and Soils from Apollo Missions." 2017, accessed August 17, 2018. www.https://curator.jsc.nasa.gov/lunar/

5 "Past Life Therapy," Encyclopedia.com, accessed November 7, 2017. https://www.encyclopedia.com/medicine/encyclopedias-almanacs-transcripts-and-maps/past-life-therapy

6 Sean Lyons, "The Science of Reincarnation." Virginia Magazine, 2013, accessed February 25, 2018. http://uvamagazine.org/articles/the_science_of_reincarnation.

7 Pim van Lommel, *Consciousness Beyond Life: The Science of the Near-Death Experience.* New York, New York, 2010.

8 Thomas S. Kuhn and Ian Hacking: *The Structure of Scientific Revolutions* (Chicago, IL: The University of Chicago Press, 2012).

## Chapter 4

1 "Carl Sagan Says the Brain Is Physical and Biological, Nothing More," The Human Truth Foundation, Accessed March 1,

2019. http://www.humantruth.info/sagan_brain_is_biologi cal_only.html.

2 "Susan Blackmore," Wikiquote, Accessed February 20,2019. https://en.wikiquote.org/wiki/Susan_Blackmore.

3 Paul Bloom, "Natural-Born Dualists," Edge, 2004, Accessed January 23,

2019. https://www.edge.org/conversation/paul_bloom-natural-born-dualists.

4 Mario Beauregard, and Denyse O'Leary, *The Spiritual Brain: A Neuroscientist's Case for the Existence of the Soul* (New York, NY: HarperOne, 2007), 289.

5 Pim van Lommel, *Consciousness Beyond Life: The Science of the Near-Death Experience* (New York, NY: HarperCollins, 2010), 166.

6 Ibid.

7 Chris Carter, *Science and the Near-Death Experience: How Consciousness Survives Death* (Rochester, NY: Inner Traditions, 2010), 236.

8 Eben Alexander, *Proof of Heaven: A Neurosurgeon's Journey into the Afterlife* (New York, NY: Simon and Schuster, 2012), 153.

9 Kenneth Ring, and Evelyn Elsaesser Valarino, *Lessons from the Light: What We Can Learn from the Near-Death Experience* (Needham MA: Moment Point, 2006), 64.

10 Kenneth Ring, and Sharon Cooper, *Mindsight: Near-Death and Out-of-Body Experiences in the Blind* (Kearney, NE: Morris Publishing, 1999) 107.

11 Van Lommel, *Consciousness Beyond Life: The Science of the Near-Death Experience*, 24.

12 Kenneth Ring, and Evelyn Elsaesser Valarino, *Lessons from the Light: What We Can Learn from the Near-Death*

*Experience*, 84.

13 Ibid., 59.

14 Ibid., 107.

15 Susan Blackmore, "First Person – into the Unknown," Dr. Susan Blackmore, 2000, Accessed March 21, 2019. https://www.susanblackmore.uk/journalism/first-person-into-the-unknown/.

16 "Tom Wilson Quotes," BrainyQuote, Accessed March 21, 2019. https://www.brainyquote.com/authors/tom_wilson.

17 Benjamin Radford, "Astral Projection: Just a Mind Trip," Live Science, 2017, Accessed June 9, 2018. https://www.livescience.com/27978-astral-projection.html.

18 Ibid.

19 "Dreaming," *Psychology Today*, Accessed December 29, 2018. https://www.psychologytoday.com/us/basics/dreaming.

20 Olaf Blanke, Stephanie Ortique, Theodor Landis, and Margitta Seeck. "Neuropsychology: Stimulating Illusory Own-Body Perceptions," *Nature* 419, no. 10 (2002): 269-70.

21 Ibid.

22 Ibid.

23 Phillip Wang, "What Are Dissociative Disorders?" American Psychiatric Association. 2016, Accessed May 28, 2018. https://www.psychiatry.org/patients-families/dissociative-disorders/what-are-dissociative-disorders.

24 Radford, "Astral Projection: Just a Mind Trip."

25 "The Vestibular System: Definition, Anatomy, and Function," Accessed May 12, 2018. https://study.com/academy/lesson/the-vestibular-system-definition-anatomy-function.html.

26 Christophe Lopez, and Maya Elziere, "Out-of-Body Experience in Vestibular Disorders – a Prospective Study of

210 Patients With Dizziness," *Cortex* 104, (2018): 193-206.

27 Ibid.

28 Alan Hale, "An Astronomer's Personal Statement on UFOs, by Alan Hale," Sacred Texts, Accessed November 6, 2018. http://www.sacred-texts.com/ufo/hale_ufo.htm.

29 Isaac Asimov, *The Roving Mind* (New York, NY: Prometheus Books, 1983), 41.

30 "Arthur C. Clark Quotes," Goodreads. Accessed August 23, 2018. https://www.goodreads.com/quotes/157737-i-m-sure-the-universe-is-full-of-intelligent-life-it-s.

31 "UFO Sightings by Astronauts," NOUFORS, 1993, Accessed May 9, 2018. http://noufors.com/ufo_sightings_by_astronauts.html.

32 Ibid.

33 Jim Lovell, "Apollo 8 Astronaut Remembers Looking Down at Earth," Smithsonian National Air and Space Museum, 2018, Accessed June 20, 2020. https://airandspace.si.edu/stories/editorial/apollo-8-astronaut-remembers-looking-down-earth.

34 Kelsey Landis, "A Successful Failure: Apollo 13 Astronaut Jim Lovell Entertains, Thrills SIUE Crowd," *The Telegraph*, 2016, Accessed July 01, 2020. https://www.thetelegraph.com/news/article/A-successful-failure-Apollo-13-astronaut-Jim-12596547.php.

35 "Edgar Mitchell," RationalWiki, Accessed May 14, 2018. https://rationalwiki.org/wiki/Edgar_Mitchell.

36 Jennifer Bayot, "Dr. John E. Mack, Psychiatrist, Dies at 74," 2004, *The New York Times*, Accessed May 29, 2018. https://www.nytimes.com/2004/09/30/us/dr-john-e-mack-psychiatrist-dies-at-74.html.

37 "Interview with John Mack, Psychiatrist, Harvard University," Ozarks Public Television, Accessed May 8, 2018. https://www.pbs.org/wgbh/nova/aliens/johnmack.html.

38 Melinda Wenner, "Belief in Reincarnation Tied to Memory Errors," NBC News, 2007, Accessed May 26, 2018. http://www.nbcnews.com/id/17982545/ns/technolog y_and_science-science/t/belief-reincarnation-tied-memory-errors/#.XFZXfM9Khje.

39 Robert Todd Carroll, "Past Life Regression," The Skeptic's Dictionary, 2015, Accessed October 2, 2018. http://skepdic.com/pastlife.html.

40 Wenner, "Belief in Reincarnation Tied to Memory Errors."

41 Ibid.

42 Gabriel Andrade, "Is Past Life Regression Therapy Ethical?," *Journal of Medical Ethics and History of Medicine* 10, no. 11 (2017): 1-8.

43 Chauncey Hollingsworth, "Altered Fate." *Chicago Tribune*, 1995, Accessed January 28, 2018. https://www.chicagotribune.com/news/ct-xpm-1995-06-21-9506210012-story.html.

44 Lisa Miller, "Remembrances of Lives Past," *The New York Times*. 2010, Accessed May 18, 2018. https://www.nytimes.com/2010/08/29/fashion/29Pa stLives.html.

45 Simon Martin, "Past Lives, Present Problems," Carol Bowman, Past Life Therapy, 1991, Accessed October 4, 2018. https://www.carolbowman.com/library/interview-dr-roger-woolger.

46 Ashok Kumar Jain, "Quantum Hypnotherapy," Accessed September 8, 2018. http://www.quantumhypnotherapy.com/workshops/quant um-hypnotherapy.aspx

47 Stephen Wagner, "Is There Evidence of Reincarnation?," ThoughtCo., 2018, Accessed September 4, 2018. https://www.thoughtco.com/reincarnation-best-evidence-2593151.

48 Jupinderjit Singh, "Can Science Uphold the Belief in

Rebirth," *The Indian Tribune*, 2002, Accessed April 5, 2018. https://www.tribuneindia.com/2002/20020622/.

## Chapter 5

1 Lorraine Davis, "A Comparison of UFO and Near-Death Experiences as Vehicles for the Evolution of Human Consciousness," *Journal of Near-Death Studies* 6, no. 4 (1988): 240-57.

2 P.M.H. Atwater, "People Are Dramatically Changed by Near-Death Experiences," Near-Death Experiences and the Afterlife, Accessed January 29, 2019. https://www.near-death.com/science/evidence/people-are-dramatically-changed-by-ndes.html.

3 Jeffrey Long, *Evidence of the Afterlife: The Science of Near-Death Experiences* (New York, NY: HarperOne, 2010), 8.

4 Pim Van Lommel, *Consciousness Beyond Life: The Science of the Near-Death Experience* (New York, NY: HarperCollins, 2010) 225

5 Ibid., 24

6 Ibid.

7 Kenneth Ring, and Sharon Cooper, *Mindsight: Near-Death and Out-of-Body Experiences in the Blind* (Kearney, NE: Morris Publishing, 1999) 112.

8 Kevin Williams, "Music and the Near-Death Experience," Near-Death Experiences and the Afterlife, Accessed January 28, 2018. https://www.near-death.com/science/research/music.html.

9 Ibid.

10 Anne Moss, "The Filter Factor," Astral Voyage, Accessed November 20, 2018. https://www.astralvoyage.com/astral-projection/filter-factor.php.

11 Ibid.

12 "Heightened Hearing When Approaching Astral Projection?," Accessed November 20, 2018.
https://www.reddit.com/r/AstralProjection/comments/8jpxho/heightened_hearing_when_approaching_astral/.

13 "Anne Moss, "The Filter Factor," Astral Voyage, Accessed November 20,

2018. https://www.astralvoyage.com/astral-projection/filter-factor.php.

14 "Sleep Paralysis, Vibrations, Spiritual Revelation, and Alien Abduction or Getting out of 4d." *Fahrusha's Weblog*, Accessed October 16,

2018. https://fahrusha.wordpress.com/2010/02/27/sleep-paralysis-vibrations-spiritual-revelation-and-alien-abduction-or-getting-out-of-4d/.

15 Ibid.

16 M. Abrams, "An Inescapable Buzz," *Discover Magazine* (1995).

17 Roger Marsh, "Florida Witnesses Alerted to Hovering UFO by Loud Humming Sound," MUFON, 2014, Accessed October 14, 2018. https://www.mufon.com/ufo-photos/previous/3.

18 Deb Belt, "25 UFO Sightings Reported in Maryland During 2018 So Far," Patch, 2018, Accessed March 25, 2018. https://patch.com/maryland/annapolis/25-ufo-sightings-reported-maryland-during-2018-so-far.

19 James M McCampbell, *Ufology: National Investigations Committee on Aerial Phenomena Chapter 4 Sounds*, Available at: https://www.nicap.org/ufology/ufochap4.htm.

20 Carol Bowman, "Experience Your Past Lives." Carol Bowman, Past Life Therapy. Accessed June 18, 2018. https://www.carolbowman.com/past-life-regression/.

21 Brian Weiss, *Messages from the Masters: Tapping into thePower of Love* (New York, NY: Grand Central Publishing 2000), 124.

22 "Secrets of Learning," Altered States, Accessed May 15, 2018. https://docs.google.com/document/d/1w5SYjQcCQBDea8y6pG7L8fz33cNoP8FG-PYEDGx9oyI/edit?ts=5ae7a36f.

23 Long, *Evidence of the Afterlife: The Science of Near-Death Experiences,* 15-16.

24 Ibid., 158.

25 "Kimberly Clark-Sharp's Near-Death Experience," Near-Death Experiences and the Afterlife, Accessed July 16, 2018. https://www.near-death.com/experiences/notable/kimberly-clark-sharp.html.

26 "Skeptical Argument: Nothing Useful Comes from NDEs," Near-Death Experiences and the Afterlife, Accessed May 23, 2018. https://www.near-death.com/index.html.

27 "The Science of Life Discovered from Lynnclaire Dennis' NDE," Near-Death Experiences and the Afterlife, Accessed January 23, 2018. https://www.near-death.com/science/articles/science-of-life-from-an-nde.html.

28 Robert Monroe, *Far Journeys.* New York, NY: DoubleDay, 1985. Available at: http://blog.hasslberger.com/docs/Far_Journeys.pdf.

29 "Alien ET Alphabet ET Script, ET Writing and Symbolism," Global ET Research, Accessed April 28, 2018. http://www.alienjigsaw.com/et-essays/Alien-Alphabet-ET-Symbols-Alien-Writings.html.

30 Louis Proud, "Aliens, Predictions & the Secret School: Decoding the Work of Whitley Strieber," New Dawn, Accessed August 18, 2018. https://www.newdawnmagazine.com/articles/aliens-predictions-the-secret-school-decoding-the-work-of-whitley-strieber.

31 Mary Rodwell, *Awakening: How Extraterrestrial Contact Can Transform Your Life* (New Mind Publishers, 2010), 243.

32 Brian Weiss, *Messages from the Masters: Tapping into the Power of Love* (New York, NY: Grand Central Publishing, 2000), 124

33 Sean Lyons, "The Science of Reincarnation," Virginia.

Accessed September 23, 2018.
http://uvamagazine.org/articles/the_science_of_reincarnati
on.

34 "Welcome to Dolores Cannon.Com," Dolores Cannon,
Accessed March 23, 2018. https://dolorescannon.com/.

35 Bruce Greyson, and Ian Stevenson, "The
Phenomenology of Near-Death Experiences," *American
Journal of Psychiatry* 137, no. 10 (1980): 1193-96.

36 Raymond Moody, *Life after Life: The Bestselling
Original Investigation That Revealed Near-Death Experiences*
(New York, NY: HarperCollins, 2001), 52-53.

37 Long, *Evidence of the Afterlife: The Science of Near-
Death Experiences,* 10.

38 "Ned Dougherty's NDE and Visions of the Future"
Near-Death Experiences and the After Life, Accessed
February 21, 2019. https://www.near-
death.com/experiences/notable/ned-dougherty.html.

39 William Buhlman, *The Secret of the Soul: Using Out-
of-Body Experiences to Understand Our True Nature* (New
York, NY: HarperCollins, 2001), 95.

40 "Interview with John Mack, Psychiatrist, Harvard
University," Ozarks Public Television, Accessed May 8,
2018. https://www.pbs.org/wgbh/nova/aliens/johnmack.ht
ml.

41 Buhlman, *The Secret of the Soul: Using Out-of-Body
Experiences to Understand Our True Nature,* 95.

42 "Life between Lives Regression," Soul Visioning,
Accessed December 14,
2018. https://www.susanwisehart.com/lifebetween.htm.

43 "My Journey into Life between Lives,"
AwakeningTimes, Accessed August 23,
2018. http://awakeningtimes.com/my-journey-into-life-
between-lives/.

44 Kock Ronald, *Introduction to Modern Spiritualism*:
Lulu, 2017, Available at: https://www.lulu.com/ (Accessed

August 27, 2018).

45 Michael Robey, "Send Love Telepathically," Psychic Tarot Oracle Reading and Courses. 2017, Accessed December 17, 2017. https://www.psychicgr.com/post/2017/01/16/send-love-telepathically.

46 "Telepathy and the Near-Death Experience," Near-Death Experiences and the Afterlife, Accessed March 8, 2018. https://www.near-death.com/science/research/telepathy.html.

47 "Near-Death Experience Research Foundation Case 1615," NDERF, Accessed September 19, 2018. https://www.nderf.org/index.htm.

48 Moody, *Life after Life: The Bestselling Original Investigation That Revealed Near-Death Experiences* 50-51.

49 "Near-Death Experience Research Foundation Case 51," NDERF, Accessed December 15, 2018. https://www.nderf.org/index.htm.

50 Sarah Williams, "Sounds You Can't Hear Can Still Hurt Your Ears," Science. 2014, Accessed January 11, 2018. https://www.sciencemag.org/news/2014/09/sounds-you-cant-hear-can-still-hurt-your-ears.

51 "The Infrasound of a Forest Elephant Rumble," The Cornell Lab of Ornithology, Accessed September 4, 2017. http://www.birds.cornell.edu/brp/elephant/cyclotis/language/infrasound.html.

52 Jenny Wade, "The Phenomenology of Near-Death Consciousness in Past-Life Regression Therapy: A Pilot Study," *Journal of Near-Death Studies* 17, no. 1 (1998): 31-53.

53 Eric Christopher, "What Past Life Regression Really Is," 2012. Accessed November 7, 2017. http://www.edgemagazine.net/2012/12/past-life-regression/.

54 "Albert Einstein Quotes," Albert Einstein Site Online 2012, Accessed March 2,

2019. http://www.alberteinsteinsite.com/quotes/einsteinqu
otes.html.

55 "Dr. P.M.H. Atwater's Near-Death Experience
Research," Near-Death Experiences and the Afterlife,
Accessed March 1, 2019. https://www.near-
death.com/science/experts/pmh-atwater.html.

56 "What Happens to the Fabric of Space-Time When an
Object Moves through It near the Speed of Light?" Gravity
Probe B: Testing Einstein's Universe, Accessed March 3,
2019.
https://einstein.stanford.edu/content/relativity/q909.html.

57 Long, *Evidence of the Afterlife: The Science of near-
Death Experiences,* 13-14.

58 Van Lommel, *Consciousness Beyond Life: The Science
of the Near-Death Experience*, xvii.

59 Albert Einstein Quotes," Albert Einstein Site Online
2012, Accessed March 2,
2019. http://www.alberteinsteinsite.com/quotes/einsteinqu
otes.html.

60 Clare Johnson, "Out-of-Body Experiences," Deep
Lucid Dreaming, Accessed November 3,
2018. http://deepluciddreaming.com/2016/02/out-of-body-
experiences/.

61 Thomas Minderle, "Astral Physics and Timespace,"
Montalk, 2008, Accessed January 14,
2019. https://montalk.net/notes/astral-physics.

62 Brent Raynes, "Reality Checking." AP Magazine, 2011,
Accessed December 26,
2018. http://www.apmagazine.info/index.php?option=com
_content&view=article&id=1152:reality-checking-june-
2018&catid=2&Itemid=44.

63 Brent Raynes, "Trying to Understand the Psychic
Mind," AP Magazine. Accessed December
27,2018. http://www.apmagazine.info/index.php?option=c
om_content&view=article&id=388:reality-checking-june-

2013&Itemid=194.

64 Starr, Michelle, "Here's How We Might Find a Wormhole, If They Existed," Science Alert, 2019, Accessed March 01, 2020.

65 "Thomas Sawyer's Near-Death Experience," Near-Death Experiences and the Afterlife, Accessed November 25, 2018. https://www.near-death.com/reincarnation/experiences/thomas-sawyer.html.

66 Brad Steiger, and Sherry Hansen Steiger, *The Gale Encyclopedia of the Unusual and Unexplained* (USA: Gale Group Inc., 2003), 69.

67 Linda Gadbois, "Reincarnation and the Soul's Journey through Time – the Holographic Nature of Reality and the Illusion of Time," Dr. Linda Gadbois, 2017, Accessed October 17, 2018. http://drlindagadbois.com/reincarnation-souls-journey-time-holographic-nature-reality-illusion-time/.

68 Atwater, "People Are Dramatically Changed by Near-Death Experiences," Near-Death Experiences and the Afterlife.

69 Long, *Evidence of the Afterlife: The Science of Near-Death Experiences,* 10.

70 Atwater, "People Are Dramatically Changed by Near-Death Experiences," Near-Death Experiences and the Afterlife.

71 Josh Lowe, "What Happens When You Die? Scientists Say Near-Death Experiences Share Elements but Vary Widely," News Week, 2017, Accessed April 6, 2018. https://www.newsweek.com/near-death-experiences-what-happens-when-you-die-643032.

72 Long, *Evidence of the Afterlife: The Science of near-Death Experiences,* 10.

73 Van Lommel, *Consciousness Beyond Life: The Science of the near-Death Experience,* 34.

74 Buhlman, *The Secret of the Soul: Using Out-of-Body Experiences to Understand Our True Nature,* 94.

75 Stuart W. Twemlow, Glenn O. Gabbard, and Fowler C. Jones, "The Out-of-Body Experience: A Phenomenological Typology Based on Questionnaire Responses," *American Journal of Psychiatry* 139, no. 4 (1982): 450-55.

76 Buhlman, *The Secret of the Soul: Using Out-of-Body Experiences to Understand Our True Nature*, 14.

77 Buhlman, *The Secret of the Soul: Using Out-of-Body Experiences to Understand Our True Nature*, 40.

77 Ibid.

78 Robert Bruce, *Astral Dynamics: The Complete Book of Out-of-Body Experiences* (Charlottesville VA: Hampton Roads Publishing Company, 2009), 256-257.

79 Buhlman, *The Secret of the Soul: Using Out-of-Body Experiences to Understand Our True Nature*, 94.

80 Susan Blackmore, "Abduction by Aliens or Sleep Paralysis?," *CSI Center For Inquiry* 22, 3 (May/June 1988), Accessed June 15, 2018. https://www.csicop.org/si/show/abduction_by_aliens _or_sleep_paralysis.

81 "Flashes of Light in Bedroom at Night," Alien Abduction Discussion Group, 2011, Accessed August 20, 2018. http://www.abduct.com/discussion/?p=815.

82 Juju Chang, and Jim Dubreuil, "Abducted by Aliens: Believers Tell Their Stories," ABC News, 2009, Accessed August 25, 2018. https://abcnews.go.com/Primetime/story?id=8330290.

83 Namina Forna, "20 People Who Think They Were Actually Abducted by Aliens Tell Their Stories," 22 Words, Accessed July 10, 2018. https://twentytwowords.com/people-who-think-they-were-actually-abducted-by-aliens-tell-their-stories/.

84 Ibid.

85 Ibid.

86 Carol Bowman, "Life between Life," Carol Bowman,

Past Life Therapy, Accessed November 28, 2018. https://www.carolbowman.com/past-life-regression/life-life/.

87 Ibid.

88 Ibid.

89 Ibid.

90 Bianca Alexander, "Learning from Your Past Lives," Conscious Living TV, Accessed January 29, 2019. http://consciouslivingtv.com/spirit/learn-past-lives.html.

91 Virginia Hummel, "What Are Orbs?," Orb Whisperer, Accessed January 20, 2019. https://orbwhisperer.com/what-are-orbs%3F.

**Chapter 6**

1 "Spontaneous Healings," Eternea, Accessed September 18, 2018. https://www.eternea.org/spontaneous_healings.aspx.

2 Caroline Myss, *Anatomy of the Spirit: The Seven Stages of Power and Healing* (New York, NY: Crown Publishing, 1996), 33.

3 David Sunfellow, "Miraculous NDE Healings." The Formula for Creating Heaven on Earth, 2019. Accessed February 15, 2019. https://the-formula.org/miraculous-nde-healings/.

4 Ibid.

5 Jeffrey Long, and Paul Perry, *Evidence of the Afterlife: The Science of Near-Death Experiences* (New York, NY: HarperOne, 2011), 188.

6 "Natalie S NDE," NDERF, Accessed July 23, 2018. https://www.nderf.org/Experiences/1natalie_s_nd e.html.

7 Long, *Evidence of the Afterlife: The Science of Near-Death Experiences*, 188.

8 "The Day My Life Changed: February 2, 2006," Anita Moorjani
Love is Not A Luxury, Accessed February 2, 2019. https://anitamoorjani.com/about-anita/near-death-experience-description/.

9 Robert Waggoner, *The Lucid Dreaming Pack: Gateway to the Inner Self* (San Francisco, CA: Moment Point Press, 2009), 155-165.

10 Preston Dennett, "UFO Healings," Preston Dennett, Accessed October 9, 2018. https://prestondennett.weebly.com/ufo-healings.html.

11 Ibid.

12 Preston Dennett, *UFO Healings: True Accounts of People Healed by Extraterrestrials* (Leland, NC: Wild Flower

Press, 1996), Accessed October 9, 2018.
https://prestondennett.weebly.com/ufo-healings.html.

13 Preston Dennett, "UFO Healings," Preston Dennett,
Accessed October 9,
2018. https://prestondennett.weebly.com/ufo-
healings.html.

14 Eric J Christopher, "Exploring the Effectiveness of
Past-Life Therapy," University of Wisconsin
Stout 2000. http://citeseerx.ist.psu.edu/viewdoc/download?
doi=10.1.1.391.2109&rep=rep1&type=pdf.

15 Ibid.

16 Heather S Friedman Rivera, *Healing the Present from
the Past, The Personal Journey of a Past-Life Researcher*
(Bloomington, IN: Balboa Press, 2012), 85-86.

17 Roger Woolger, "Past Life Therapy, Trauma Release
and the Body," Carol Bowman, Past Life Therapy. Accessed
February 6,
2019. https://www.carolbowman.com/library/past-life-
therapy-trauma-release-body.

18 "Phobias," Medline Plus, Accessed November 21
2018. https://medlineplus.gov/phobias.html.

19 Lisa Fritscher, "How Many People Have Phobias?"
verywellmind, 2018. Accessed March 12,
2019. https://www.verywellmind.com/prevalence-of-
phobias-in-the-united-states-2671912.

20 Brian Weiss, "Many Lives, Many Masters," 1988,
Accessed February 6,
2019. http://siqv.weebly.com/uploads/2/3/5/1/23516512/m
any_lives_many_masters_-_brian_weiss.pdf.

21 Rivera, *Healing the Present from the Past, The
Personal Journey of a Past-Life Researcher*, 53

22 "Past Life Regression (PLR) #3: Healing Emotional
Trauma with PLR," Daniel Olexa Hyponotherapy. 2016,
Accessed December 18, 2018.
https://www.danielolexa.com/blog/healing-emotional-

trauma-with-plr.

23 Judith Pennington, "P.M.H. Atwater and the Near-Death Experience," Institute for the Awakened Mind, 2017. Accessed June 5, 2018. http://iamevents.tilda.ws/atwaternde.

## Chapter 7

1 "About STES," American Center for the Integration of Spiritually Transformative Experiences, Accessed March 13, 2019. https://aciste.org/about-stes/.

2 Brian Weiss, *Many Lives, Many Masters* (New York, NY: Fireside, 1988), 208.

3 Pim Van Lommel, *Consciousness Beyond Life: The Science of the Near-Death Experience* (New York, NY: HarperCollins, 2010), 47.

4 I bid.

5 P.M.H. Atwater, "People Are Dramatically Changed by Near-Death Experiences," Near-Death Experiences and the Afterlife, Accessed January 29, 2019. https://www.near-death.com/science/evidence/people-are-dramatically-changed-by-ndes.html.

6 Ibid.

7 Van Lommel, *Consciousness Beyond Life: The Science of the Near-Death Experience*, 68.

8 P.M.H. Atwater, *Physiological Aftereffects of Near-Death States*, Beyond the Light, (New York, NY: Avon Books, 1994), Available at: http://pmhatwater.hypermart.net/resources/PDFs/Articles/C-Aftereffects-Percentages-3.pdf

9 Ibid.

10 Van Lommel, *Consciousness Beyond Life: The Science of the Near-Death Experience*, 68.

11 Jeffrey Long, and Paul Perry, *Evidence of the Afterlife: The Science of Near-Death Experiences* (New York, NY: HarperOne, 2010) 181.

12 Van Lommel, *Consciousness Beyond Life: The Science of the Near-Death Experience*, 68.

13 Ibid.

14 Ibid.

15 Ibid.

16 Atwater, *Physiological Aftereffects of Near-Death*

*States.*

17 Ibid.

18 Ibid.

19 Ibid.

20 Ibid.

21 Ibid.

22 Van Lommel, *Consciousness Beyond Life: The Science of the Near-Death Experience*, 61.

23 Atwater, *Physiological Aftereffects of Near-Death States.*

24 Ibid.

25 Van Lommel, *Consciousness Beyond Life: The Science of the Near-Death Experience*, 61.

26 Atwater, *Physiological Aftereffects of Near-Death States.*

27 Van Lommel, *Consciousness Beyond Life: The Science of the Near-Death Experience*, 61.

28 Atwater, *Physiological Aftereffects of Near-Death States.*

29 Ibid.

30 P.M.H. Atwater, "Near-Death States: The Pattern of Aftereffects." (4. http://pmhatwater.hypermart.net/resources/PDFs/Articles/Hannover-NDE-Aftereffect.pdf. (Accessed October 4, 2018), 4.

31 Ibid.

32 William Buhlman, "Survey Results." Astralinfo.org, Accessed July 18, 2018. http://www.astralinfo.org/survey-results/.

33 Stuart W. Twemlow, Glenn O. Gabbard, and Fowler C. Jones, "The Out-of-Body Experience: A Phenomenological Typology Based on Questionnaire Responses," *American Journal of Psychiatry* 139, no. 4 (1982): 450-455.

34 Ibid.

35 Ibid.

36 Ibid.

37 Ibid.

38 Ibid.

39 Ibid.

40 Ibid.

41 Carlos S Alvardo, and Nancy L. Zingron, "Exploring the Factors Related to the After-Effects of Out-of-Body Experiences," *Journal of the Society for Psychical Research* 67, no. 872 (2003): 161-83.

42 Ibid.

43 Ibid.

44 Ibid.

45 Twemlow, Glenn O. Gabbard, and Fowler C. Jones, "The Out-of-Body Experience: A Phenomenological Typology Based on Questionnaire Responses, 450-455.

46 Ibid.

47 Ibid.

48 Ibid.

49 Ibid.

50 Ibid.

51 Alexander DeFoe, "How Should Therapists Respond to Client Accounts of Out-of-Body Experience?," *International Journal of Transpersonal Studies* 31, no. 1 (2012): 75-82.

52 Stuart Twemlow, "Clinical Approaches to the Out-of-Body Experience," *Journal of Near-Death Studies* 8, no. 1 (1989): 29-43.

53 DeFoe, "How Should Therapists Respond to Client Accounts of Out-of-Body Experience?" 75-82.

54 Kathleen Marden, "The Marden-Stoner Study on Commonalities among UFO Abduction Experiencers," *Kathleen Marden*, Accessed February 5, 2019. http://www.kathleen-marden.com/resources/The%20Marden-Stoner%20Study%20on%20Commonalities%20Among%2

oAbduction%20Experiencers.pdf.

55 Ibid.

56 Ibid.

57 Ibid.

58 Kenneth Ring, "The Omega Project: Near-Death Experiences, UFO Encounters, and Mind at Large," The Intuitive-Connections Network, Accessed June 24, 2018. http://www.intuitive-connections.net/2008/book-omegaproject.htm.

59 Roberta Colasanti, "Integrating Extraordinary Experiences," John E. Mack Institute, 1999, Accessed September 26, 2017. http://johnemackinstitute.org/1999/04/integrating-extraordinary-experiences/.

60 Eric J. Christopher, *Exploring the Effectiveness of Past-Life Therapy*, The Graduate College University of Wisconsin - Stout, 2000. http://citeseerx.ist.psu.edu/viewdoc/download?doi=10.1.1.391.2109&rep=rep1&type=pdf.

61 Ibid.

62 Ibid.

63 Heather Rivera, "Healing the Present from the Past." PhD Dissertation. Accessed February 5, 2019.

64 Ibid.

65 Ibid.

66 Ibid.

67 Ibid.

68 Ibid.

69 Ibid.

70 Ibid.

71 Ibid.

72 Ibid.

73 Ibid.

74 Jim Tucker, "Advice to Parents of Children Who Are Spontaneously Recalling Past Life Memories," Division of

Perceptual Studies, Accessed May 17,
2018. https://med.virginia.edu/perceptual-
studies/resources/advice-to-parents-of-children-who-are-
spontaneously-recalling-past-life-memories/.

75 Ibid.

76 Sean Lyons, "The Science of Reincarnation," Virginia
Magazine, 2013, Accessed February 25,
2018. http://uvamagazine.org/articles/the_science_of_reinc
arnation.

77 Tucker, "Advice to Parents of Children Who Are
Spontaneously Recalling Past Life Memories.

78 Kalki Mahavatar, "Consciousness and Vibrational
Frequencies – Resonating in Oneness with Cosmos
Supersoul," 2015. Accessed November 18,
2018. https://www.linkedin.com/pulse/consciousness-
vibrational-frequencies-resonating-oneness-priya/.

79 David Talbott, and Wallace Thornhill, "Excerpts from
Thunderbolts of the Gods," The Thunderbolts Project,
Accessed July 29,
2017. http://www.thunderbolts.info/wp/resources/thunder
bolts-of-the-gods-excerpts/.

80 Richard Bonenfant, "A Study of the Relationship
between Abduction Experiences and Unusual
Electromagnetic after-Effects: A Summary Review, Report
and Discussion," The-Electro-Magnetic-E-M-Effects-Project,
2016, Accessed June 3,
2017. http://www.alienjigsaw.com/anomalies/Bonenfant-A-
Study-of-the-Relationship-Between-%0AAbductees-and-
Elecromagnetic-After-effects.html.

81 P.M.H. Atwater, "Aftereffects of Near-Death States,"
The Website of P.M.H. Atwater, Accessed February 4,
2018. http://pmhatwater.hypermart.net/page10/NDEAfter.
html.

82 Bruce Greyson, Mitchell Leister, Lee Kinsey, Steve
Alsum, and Glen Fox, "Electromagnetic Phenomena

Reported by Near-Death Experiencers," *Journal of Near-Death* 33, no. 4 (2015): 213-27.

83 Buhlman, "Survey Results."

84 William Buhlman, "Vibrational State." Astralinfo.org, Accessed August 7,

2018. http://www.astralinfo.org/?s=vibration.

85 Bonenfant, "A Study of the Relationship between Abduction Experiences and Unusual Electromagnetic After-Effects: A Summary Review, Report and Discussion."

86 Ibid.

87 Marden, "The Marden-Stoner Study on Commonalities among UFO Abduction Experiencers."

88 Edwige Bingue, *You're Not Crazy, You're Awakening: Journey to Discovering Your Soul Purpose* (Trafford Publishing, 2013), 99.

## Chapter 8

1 "Knowledge Quotes," Good Reads, Accessed January 23, 2019. https://www.goodreads.com/quotes/tag/knowledge.

2 James M Ambury, "Socrates (469–399 B.C.E.)," Internet Encyclopedia of Philosophy, Accessed September 30, 2018. https://www.iep.utm.edu/socrates/.

3 Ibid.

## Chapter 9

1 Socrates Quotes," Goodreads, Accessed March 2, 2019. https://www.goodreads.com/quotes/113638-the-greatest-blessing-granted-to-mankind-come-by-way-of.

2 "Thomas Szasz," Wikiquote, Accessed March 15, 2019. https://en.wikiquote.org/wiki/Thomas_Szasz.

3 Paul Levy, "Psychiatry Almost Drove Me Crazy," Awaken in the Dream, Accessed January 11, 2019. https://www.awakeninthedream.com/articles/psychiatry-almost-drove-me-crazy-2.

4 Satwant K. Pasricha, "Relevance of Para-Psychology in Psychiatric Practice," *Indian Journal of Psychiatry* 53, no. 1 (2011): 4-8.

5 "Center for Consciousness Studies," The University of Arizona, Accessed January 11, 2019. https://directory.arizona.edu/centers/center-consciousness-studies.

6 "Top Universities in the World 2019," TopUniversities, 2019, Accessed February 10, 2019. https://www.topuniversities.com/university-rankings-articles/world-university-rankings/top-universities-world-2019.

7 Daryl J. Bem, "Feeling the Future: Experimental Evidence for Anomalous Retroactive Influences on Cognition and Affect," *Journal of Personality and Social Psychology* 100, no. 3 (2011): 407-25.

8 "World University Rankings," TopUniversities, 2019, Accessed February 10, 2019. https://www.topuniversities.com/university-rankings/world-university-rankings/2019.

9 Benedict Cary, "A Princeton Lab on ESP Plans to Close Its Doors," The New York Times, 2007, Accessed March 30, 2019. https://www.nytimes.com/2007/02/10/science/10princeton.html.

10 Paul Seaburn, "UFO Studies to Be Taught at Major

University," Mysterious Universe, 2018, Accessed January 15, 2019. https://mysteriousuniverse.org/2018/12/ufo-studies-to-be-taught-at-duke-university/.

11 Ibid.

12 "John Carew Eccles," Wikiquote, Accessed March 2, 2019. https://en.wikiquote.org/wiki/John_Carew_Eccles.

## Chapter 10

1 "Why Near-Death Experiences Are Not Hallucinations," Near-Death Experiences and the Afterlife, Accessed January 28, 2019. https://www.near-death.com/science/hallucinations/why-ndes-are-not-hallucinations.html.

2 "Science and the Near-Death Experience," Near-Death Experiences and the Afterlife, Accessed March 18, 2019. https://www.near-death.com/science/research/science.html.

3 "Who Is Saying It? – Quotes," Youareme.com, Accessed February 16, 2019. http://www.youareme.com/quotes.html.

4 Ibid.

5 Larry Dossey, "Why Consciousness Is Not in the Brain," Pathways to Family Wellness, Accessed March 4, 2019. http://pathwaystofamilywellness.org/New-Edge-Science/why-consciousness-is-not-the-brain.html.

6 "A Quantum Theory of Consciousness," Huffington Post, 2017, Accessed January 30, 2019. https://www.huffingtonpost.com/entry/a-quantum-theory-of-consciousness_us_596fb782e4b04dcf308d29bb.

7 Pim van Lommel, "Near-Death Experiences: The Experience of the Self as Real and Not as an Illusion," *New York Academy of Sciences* 1234 (2011): 19-28. https://www.ncbi.nlm.nih.gov/pubmed/21988246.

8 "This Is Exactly Why Nikola Tesla Told Us to Study the 'Non Physical,'" Cosmic Scientist, 2016, Accessed January 20, 2019. http://www.cosmicscientist.com/this-is-exactly-why-nikola-tesla-told-us-to-study-the-non-physical/.

9 "Crows Quotes," Goodreads, Accessed March 11, 2019. https://www.goodreads.com/quotes/tag/crows.

10 "William Lawrence Bragg," Goodreads, Accessed March 12, 2019. https://www.goodreads.com/quotes/166826-the-important-thing-in-science-is-not-so-much-to.

# Bibliography
## For Images

1-1 Original artwork by Deanna Miller. *Illustration of Egyptian hieroglyphic (@2,000-3,000 BCE, depicting ancient astronauts).*

1-2 Original artwork by Deanna Miller. *Illustrative representation of non-local consciousness.*

2-1 Original artwork by Deanna Miller. *Illustration of Albert Einstein-Physicist and "father" of The Theory of Relativity.*

2-2 Original chart created by Lynn Miller. *Myers-Briggs Type Indicator. (MBTI). Assessment designed to pinpoint psychological preferences in how individuals perceive the world and make decisions.*

2-3 Original graph by Deanna Miller. *Gregorc Mind Styles Four Quadrant Model. Assessment which determines the four preferred learning styles based on perceptual and processing modes.*

3-1 Original artwork by Deanna Miller. *Illustration of out-of-body experience.*

4-1 Original artwork by Deanna Miller. *Illustration of skepticism.*

5-1 Original artwork by Deanna Miller. *Unidentified aerial phenomena abduction.*

5-2 "Flower of Life Sacred Geometry." Pixabay, accessed February, 05, 2020, https://pixabay.com/vectors/flower-of-life-sacred-geometry-1601163/.

5-3 "Fibonacci Geometry Mathematics." Pixabay, accessed February, 12, 2020, https://pixabay.com/illustrations/fibonacci-geometry-mathematics-1079783/.

5-4 Original artwork by Deanna Miller. *Representative of non-linear time experienced in extraordinary phenomena.*

5-5 Permission granted from Barbara Mango, personal photography collection. Wormhole, Miami FL, 2017.

5-6 Permission granted from Barbara Mango, personal photography collection. Lights in the Sky, Miami FL, 2017.

5-7 Permission granted from Tom Dongo, personal photography collection. *Being of light.*

5-8 Original artwork by Deanna Miller. *String Theory. Representation of multidimensional reality.*

5-9 Original artwork by Deanna Miller. *Representation of vortexes experienced during OBEs, while traveling from one dimension to another.*

5-10 Permission granted from Tom Dongo. *Merging Dimensions: The Opening Portals of Sedona. Photograph of multiple orbs.* Flagstaff, AZ: Light Technology Publishing, 1995.

6-1 Original artwork by Deanna Miller. *Representation of energetic healing.*

7-1 Original artwork by Deanna Miller. *Blocking EM waves.*

7-2 Original artwork by Deanna Miller. *Subtle energetic bodies.*

# Bibliography

"A Quantum Theory of Consciousness." Huffington Post, 2017, accessed January 30, 2019, https://www.huffingtonpost.com/entry/a-quantum-theory-of-consciousness_us_596fb782e4b04dcf308d29bb.

"About STES." American Center for the Integration of Spiritually Transformative Experiences, accessed March 13, 2019, https://aciste.org/about-stes/.

Abrams, M. "An Inescapable Buzz." *Discover Magazine* (1995).

"Aftereffects of Near-Death States." The Website of P.M.H. Atwater, accessed February 4, 2018, http://pmhatwater.hypermart.net/page10/NDEAfter.html.

"Albert Einstein Quotes." Albert Einstein Site Online, 2012, accessed March 2, 2019, http://www.alberteinsteinsite.com/quotes/einsteinquotes.html.

Alexander, Bianca. "What I Learned from My Past Life Regression." Conscious Living TV, accessed January 29, 2019, http://consciouslivingtv.com/spirit/learn-past-lives.html.

Alexander, Eben. *Proof of Heaven: a Neurosurgeon's Journey into the Afterlife.* New York, NY: Simon and Schuster, 2012.

"Alien ET Alphabet ET Script, ET Writing and Symbolism." Global ET Research, accessed April 28, 2018, http://www.alienjigsaw.com/et-essays/Alien-Alphabet-ET-Symbols-Alien-Writings.html.

Alvarado, Carlos S., and Nancy L. Zingron. "Exploring the Factors Related to the after-Effects of out-of-Body Experiences." *Journal of the Society for Psychical Research* 67, no. 872 (2003): 161-83.

Andrade, Gabriel. "Is Past Life Regression Therapy

Ethical?". *Journal of Medical Ethics and History of Medicine* 10, no. 11 (2017): 1-8. https://www.ncbi.nlm.nih.gov/pmc/articles/PMC5797677/.

"Anomalous." Dictionary by Merriam-Webster, accessed December 10, 2018, https://www.merriam-webster.com/.

"Arthur C. Clark Quotes." Goodreads, accessed August 23, 2018, https://www.goodreads.com/quotes/157737-i-m-sure-the-universe-is-full-of-intelligent-life-it-s.

Asimov, Isaac. *The Roving Mind*. New York, NY: Prometheus Books, 1983.

"Astral Physics and Timespace." Montalk, 2008, accessed January 14, 2019, https://montalk.net/notes/351/astral-physics-and-timespace.

A Study of the Relationship between Abduction Experiences and Unusual Electromagnetic After-Effects: A Summary Review, Report and Discussion." The-Electro-Magnetic-E-M-Effects-Project, 2016, accessed June 3, 2017, http://www.alienjigsaw.com/anomalies/Bonenfant-A-Study-of-the-Relationship-Between-%0AAbductees-and-Elecromagnetic-After-effects.html.

Atwater, P. M. H. "Near-Death States: The Pattern of Aftereffects." Accessed October 4, 2018. http://pmhatwater.hypermart.net/resources/PDFs/Articles/Hannover-NDE-Aftereffect.pdf.

Atwater, P.M.H. "Physiological Aftereffects of Near-Death States." In *Beyond the Light* New York, NY: Avon Books, 1994http://pmhatwater.hypermart.net/resources/PDFs/Articles/C-Aftereffects-Percentages-3.pdf.

Bayot, Jennifer. "Dr. John E. Mack, Psychiatrist, Dies at 74," 2004, accessed May 29, 2018, https://www.nytimes.com/2004/09/30/us/dr-john-e-mack-psychiatrist-dies-at-74.html.

Beauregard, Mario, and Denyse O'Leary. *The Spiritual Brain: A Neuroscientist's Case for the Existence of the*

*Soul.* New York, NY: HarperOne, 2007.

"Being Left-Handed." Left Handed Facts, accessed June 5, 2018, https://www.lefthandersday.com/tour/being-left-handed#.XGiiDaJKjIV.

Belt, Deb. "25 UFO Sightings Reported in Maryland During 2018 So Far." Patch, 2018, accessed March 25, 2018, https://patch.com/maryland/annapolis/25-ufo-sightings-reported-maryland-during-2018-so-far.

Bem, Daryl J. "Feeling the Future: Experimental Evidence for Anomalous Retroactive Influences on Cognition and Affect." *Journal of Personality and Social Psychology* 100, no. 3 (2011): 407-25.

Bingue, Edwige. *You're Not Crazy, You're Awakening: Journey to Discovering Your Soul Purpose.* Trafford Publishing, 2013.

Blackmore, Susan. "Abduction by Aliens or Sleep Paralysis?" *Skeptical Inquirer* 22, no. 3. (May/June 1988). Accessed June 15, 2018. https://www.csicop.org/si/show/abduction_by_aliens_or_sleep_paralysis.

Blackmore, Susan. "First Person – into the Unknown." Dr Susan Blackmore, 2000, accessed March 21, 2019, https://www.susanblackmore.uk/journalism/first-person-into-the-unknown/.

Blanke, Olaf, Stephanie Ortique, Theodor Landis, and Margitta Seeck. "Neuropsychology: Stimulating Illusory Own-Body Perceptions." *Nature* 419, no. 10 (2002): 269-70.

Bloom, Paul. "Natural-Born Dualists." Edge, 2004, accessed January 23, 2019, https://www.edge.org/conversation/paul_bloom-natural-born-dualists.

Bowman, Carol. "Experience Your Past Lives." Carol Bowman, Past Life Therapy, accessed June 18, 2018, https://www.carolbowman.com/past-life-regression/.

Bruce, Robert. *Astral Dynamics: The Complete Book of*

*Out-of-Body Experiences*. Charlottesville VA: Hampton Roads Publishing Company, 2009.

Buhlman, William. *The Secret of the Soul: Using Out-of-Body Experiences to Understand Our True Nature*. New York, NY: HarperCollins, 2001.

Carey, Benedict. "A Princeton Lab on ESP Plans to Close Its Doors." The New York Times, 2007, accessed March 30, 2019, https://www.nytimes.com/2007/02/10/science/10princeton.html.

Carter, Chris. *Science and the Near-Death Experience: How Consciousness Survives Death*. Rochester, NY: Inner Traditions, 2010.

"Center for Consciousness Studies." The University of Arizona, accessed January 11, 2019, https://directory.arizona.edu/centers/center-consciousness-studies.

Chang, Juju, and Jim Dubreuil. "Abducted by Aliens: Believers Tell Their Stories." ABC News, 2009, accessed August 25, 2018, https://abcnews.go.com/Primetime/story?id=8330290.

Christopher, Eric J. "What Past Life Regression Really Is." 2012, The Edge, accessed November 7, 2017, http://www.edgemagazine.net/2012/12/past-life-regression/.

Christopher, Eric J. "Exploring the Effectiveness of Past-Life Therapy." 32-33. The Graduate College University of Wisconsin - Stout, 2000. Research Paper.

Colasanti, Roberta L. "Integrating Extraordinary Experiences." John E. Mack Institute, 1999, accessed September 26, 2017, http://johnemackinstitute.org/1999/04/integrating-extraordinary-experiences/.

"Commonly Reported Phenomena Associated with Out-of-Body Experiences." Astralinf.org, accessed December 31,

2018, http://www.astralinfo.org/survey-results/.

"Consciousness Science: A Brief Introduction."
International Academy of Consciousness, accessed June 10,
2018, https://www.iacworld.org/consciousness-science-introduction/.

Crabtree, Vexen. "Carl Sagan Says the Brain Is Physical
and Biological, Nothing More." The Human Truth
Foundation, accessed March 1,
2019, http://www.humantruth.info/sagan_brain_is_biological_only.html.

"Crows Quotes." Goodreads, accessed March 11,
2019, https://www.goodreads.com/quotes/tag/crows.

Davis, Lorraine. "A Comparison of UFO and Near-Death
Experiences as Vehicles for the Evolution of Human
Consciousness." Journal of Near-Death Studies 6, no. 4
(1988): 240-57.

"Definition of Unidentified Aerial Phenomena, UAP."
National Aviation Reporting Center on Anomalous
Phenomena, accessed January 31,
2019, http://www.narcap.org/Blog

DeFoe, Alexander. "How Should Therapists Respond to
Client Accounts of out-of-Body Experience?." International
Journal of Transpersonal Studies 31, no. 1 (2012): 75-82.

Dennett, Daniel. "Consciousness Explained." Boston,
MA: Little Brown 1991.

Dennett, Preston. UFO Healings: True Accounts of People
Healed by Extraterrestrials. Leland, NC: Wild Flower Press,
1996.

Dennett, Preston. "UFO Healings." Accessed October 9,
2018, https://prestondennett.weebly.com/ufo-healings.html.

"Dissociative Disorders." National Alliance on Mental
Illness, accessed December 26,
2018, https://www.nami.org/learn-more/mental-health-conditions/dissociative-disorders.

Dossey, Larry. "Why Consciousness Is Not in the Brain." Pathways to Family Wellness, accessed March 4, 2019, http://pathwaysoffamilywellness.org/New-Edge-Science/why-consciousness-is-not-the-brain.html.

"Dr. Richard Eby's NDE and the Second Coming of Christ." Near-Death Experiences and the Afterlife, accessed November 9, 2018, https://www.near-death.com/science/articles/richard-eby-and-secomd-coming-of-christ.html.

"Dr. P.M.H. Atwater's Near-Death Experience Research." Near-Death Experiences and the Afterlife, accessed March 1, 2019, https://www.near-death.com/science/experts/pmh-atwater.html.

"Dreaming." Psychology Today, accessed December 29, 2018, https://www.psychologytoday.com/us/basics/dreaming.

"Edgar Mitchell." RationalWiki, accessed May 14, 2018, https://rationalwiki.org/wiki/Edgar_Mitchell.

"Flashes of Light in Bedroom at Night." Alien Abduction Discussion Group, 2011, accessed August 20, 2018, http://www.abduct.com/discussion/?p=815.

Forna, Namina. "20 People Who Think They Were Actually Abducted by Aliens Tell Their Stories." 22 Words, accessed July 10, 2018, https://twentytwowords.com/people-who-think-they-were-actually-abducted-by-aliens-tell-their-stories/.

Fritscher, Lisa. "How Many People Have Phobias?" verywellmind, 2018, accessed March 12, 2019, https://www.verywellmind.com/prevalence-of-phobias-in-the-united-states-2671912.

Gadbois, Linda. "Reincarnation and the Soul's Journey through Time – the Holographic Nature of Reality and the Illusion of Time." Dr. Linda Gadbois, 2017, accessed October 17, 2018, http://drlindagadbois.com/reincarnation-souls-journey-time-holographic-nature-reality-illusion-time/.

Greyson, Bruce, Mitchell Leister, Lee Kinsey, Steve Alsum, and Glen Fox. "Electromagnetic Phenomena Reported by Near-Death Experiencers." *Journal of Near-Death* 33, no. 4 (2015): 213-27.

Greyson, Bruce, and Ian Stevenson. "The Phenomenology of Near-Death Experiences." *American Journal of Psychiatry* 137, no. 10 (1980): 1193-96.

Hale, Alan. "An Astronomer's Personal Statement on UFOs." Sacred Texts, accessed November 6, 2018, http://www.sacred-texts.com/ufo/hale_ufo.htm.

"Heightened Hearing When Approaching Astral Projection?" r/AstralProjection, accessed November 20, 2018, https://www.reddit.com/r/AstralProjection/comments/8jpxho/heightened_hearing_when_approaching_astral/.

Hollingsworth, Chauncey. "Altered Fate." *Chicago Tribune*, 1995, accessed January 28, 2018, https://www.chicagotribune.com/news/ct-xpm-1995-06-21-9506210012-story.html.

"Interview with John Mack, Psychiatrist, Harvard University." Ozarks Public Television, accessed May 8, 2018, https://www.pbs.org/wgbh/nova/aliens/johnmack.html.

"Interview with John Mack, Psychiatrist, Harvard University," accessed July 15, 2018, https://www.pbs.org/wgbh/nova/aliens/johnmack.html.

Irwin, Harvey J. "Fantasy Proneness and Paranormal Beliefs." *Psychological Reports* 66, no. 2 (1990): 655-58.

Jain, Ashok K. "Quantum Hypnotherapy ," accessed September 8, 2018, http://www.quantumhypnotherapy.com/workshops/quantum-hypnotherapy.aspx.

"John Carew Eccles." Wikiquote, accessed March 2, 2019, https://en.wikiquote.org/wiki/John_Carew_Eccles.

Johnson, Clare. "Out-of-Body Experiences." Deep Lucid

Dreaming, accessed November 3, 2018, http://deepluciddreaming.com/2016/02/out-of-body-experiences/.

"Kimberly Clark-Sharp's Near-Death Experience." Near-Death Experiences and the Afterlife, accessed July 16, 2018, https://www.near-death.com/experiences/notable/kimberly-clark-sharp.html.

King, Julie. "What Is Your Dominant Thinking Style?" CanadaOne, 2014, accessed May 5, 2018, https://www.canadaone.com/ezine/2013/what_are_your_dominant_thinking_styles.html.

"Knowledge Quotes." Good Reads, accessed January 23, 2019, https://www.goodreads.com/quotes/tag/knowledge.

Kock, Ronald. *Introduction to Modern Spiritualism*. Lulu, 2017. https://www.lulu.com/.

Kuhn, Thomas S., and Ian Hacking. *The Structure of Scientific Revolutions*. Fourth edition. ed. Chicago, IL: The University of Chicago Press, 2012.

Lake, James. "The Evolution of a Predisposition for the near-Death Experience: Implications for Non-Local Consciousness." *Journal of Nonlocality* 5, no. 1 (2017): 1-17.

Landis, Kelsey. "A Successful Failure: Apollo 13 Astronaut Jim Lovell Entertains, Thrills SIUE Crowd," *The Telegraph*, 2016, accessed July 01, 2020, https://www.thetelegraph.com/news/article/A-successful-failure-Apollo-13-astronaut-Jim-12596547.php.

Levy, Paul. "Psychiatry Almost Drove Me Crazy." Awaken in the Dream, accessed January 11 2019, https://www.awakeninthedream.com/articles/psychiatry-almost-drove-me-crazy-2.

"Life between Life." Carol Bowman, Past Life Therapy, accessed November 28, 2018, https://www.carolbowman.com/past-life-regression/life-life/.

"Life between Lives Regression." Soul Visioning,

accessed December 14, 2018, https://www.susanwisehart.com/lifebetween.htm.

Long, Jeffrey. *Evidence of the Afterlife: The Science of Near-Death Experiences*. New York, NY: HarperOne, 2010.

Long, Jeffrey. "Near Death Experience Overview." Near-Death Experience Research Foundation, 1999, accessed August 15, 2018, https://www.nderf.org/NDERF/Articles/NDE%20Overview.htm.

Long, Jeffrey, and Paul Perry. *Evidence of the Afterlife: The Science of Near-Death Experiences*. New York, NY: HarperOne, 2010.

Lopez, Christophe, and Maya Elziere. "Out-of-Body Experience in Vestibular Disorders – a Prospective Study of 210 Patients with Dizziness." *Cortex* 104 (2018): 193-206.

Lovell, Jim. "Apollo 8 Astronaut Remembers Looking Down at Earth," Smithsonian National Air and Space Museum, 2018, Accessed June 20, 2020, https://airandspace.si.edu/stories/editorial/apollo-8-astronaut-remembers-looking-down-earth.

Lowe, Josh. "What Happens When You Die? Scientists Say Near-Death Experiences Share Elements but Vary Widely." *Newsweek*, 2017, accessed April 6, 2018, https://www.newsweek.com/near-death-experiences-what-happens-when-you-die-643032.

"Lunar Rocks and Soils from Apollo Missions." NASA, 2017, accessed April 18, 2018, www.https://curator.jsc.nasa.gov/lunar/.

Lyons, Sean. "The Science of Reincarnation." *University of Virginia Magazine*, 2013, accessed February 25, 2018, and September 23, 2018, http://uvamagazine.org/articles/the_science_of_reincarnation.

Mack, John E. *Abduction: Human Encounters with Aliens*. New York, NY: Simon and Schuster, 1994.

Mahavatar, Kalki. "Consciousness and Vibrational Frequencies – Resonating in Oneness with Cosmos Supersoul." 2015, accessed November 18, 2018, https://www.linkedin.com/pulse/consciousness-vibrational-frequencies-resonating-oneness-priya/.

"Manifesto for a Post-Materialist Science." 2014, accessed June 14, 2018, http://opensciences.org/about/manifesto-for-a-post-materialist-science.

Marden, Kathleen. "The Marden-Stoner Study on Commonalities among UFO Abduction Experiencers." 4-13. Accessed February 5, 2019. http://www.kathleen-marden.com/resources/The%20Marden-Stoner%20Study%20on%20Commonalities%20Among%20Abduction%20Experiencers.pdf.

Marsh, Roger. "Florida Witnesses Alerted to Hovering UFO by Loud Humming Sound." MUFON, 2014, accessed October 14, 2018, https://www.mufon.com/ufo-photos/previous/3.

Martin, Simon. "Past Lives, Present Problems." Carol Bowman, Past Life Therapy, 1991, accessed October 4, 2018, https://www.carolbowman.com/library/interview-dr-roger-woolger.

McCampbell, James M. *Ufology: National InvestigationsCommittee on Aerial Phenomena.* https://www.nicap.org/ufology/ufochap4.htm.

Miller, Lisa. "Remembrances of Lives Past." *The New York Times*, 2010, accessed May 18, 2018, https://www.nytimes.com/2010/08/29/fashion/29PastLives.html.

"Mind Styles-Anthony Gregorc." accessed November 13, 2018, http://web.cortland.edu/andersmd/learning/Gregorc.htm.

Monroe, Robert. *Far Journeys.* New York, NY: DoubleDay, 1985.

http://blog.hasslberger.com/docs/Far_Journeys.pdf.

Moody, Raymond. *Life after Life: The Bestselling Original Investigation That Revealed "Near-Death Experiences.* New York, NY: HarperCollins, 2001.

"Music and the Near-Death Experience." Near-Death Experiences and the Afterlife, accessed January 28, 2018 and June 23, 2018, https://www.near-death.com/science/research/music.html.

"My Journey into Life between Lives." AwakeningTimes, accessed August 23, 2018, http://awakeningtimes.com/my-journey-into-life-between-lives/.

Myss, Caroline. *Anatomy of the Spirit: The Seven Stages of Power and Healing.* New York, NY: Crown Publishing, 1996.

"Natalie S Nde." NDERF, accessed July 23, 2018, https://www.nderf.org/Experiences/1natalie_s_nde.html.

"Near-Death Experience Research Foundation Case 51." NDERF, accessed December 15, 2018, https://www.nderf.org/index.htm.

"Near-Death Experience Research Foundation Case 1615." NDERF, accessed September 19, 2018, https://www.nderf.org/index.htm.

"Ned Dougherty's NDE and Visions of the Future." Near-Death Experiences and the After Life, accessed February 21, 2019, https://www.near-death.com/experiences/notable/ned-dougherty.html.

Nickell, Joe. "A Study of Fantasy Proneness in the Thirteen Cases of Alleged Encounters in John Mack's Abduction." *Investigative Files* 20, no. 3. (May/June 1996). https://www.csicop.org/si/show/study_of_fantasy_proneness_in_the_thirteen_cases_of_alleged_encounters_in_j.

"Overwhelmed by the World?" Inner World, accessed December 29,

2018, http://www.aislingmagazine.com/aislingmagazine/ar ticles/TAM28/Overwhelmed.html.

Pasricha, Satwant K. "Relevance of Para-Psychology in Psychiatric Practice." *Indian Journal of Psychiatry* 53, no. 1 (2011): 4-8.

"Past Life Regression." The Skeptic's Dictionary, 2015, accessed October 2, 2018, http://skepdic.com/pastlife.html.

"Past Life Regression (PLR) #3: Healing Emotional Trauma with PLR." Daniel Olexa Hypnotherapy, 2016, accessed December 18, 2018, https://www.danielolexa.com/blog/healing-emotional-trauma-with-plr.

"Past Life Therapy." Encyclopedia.com, accessed November 7, 2017, https://www.encyclopedia.com/medicine/encyclopedi as-almanacs-transcripts-and-maps/past-life-therapy.

"People Are Dramatically Changed by Near-Death Experiences." Near-Death Experiences and the Afterlife, accessed January 29, 2019, https://www.near-death.com/science/evidence/people-are-dramatically-changed-by-ndes.html.

"P.M.H. Atwater and the Near-Death Experience." Institute For the Awakened Mind, 2017, accessed June 5, 2018, http://iamevents.tilda.ws/atwaternde.

"Phobias." Medline Plus, accessed November 21, 2018, https://medlineplus.gov/phobias.html.

Powers, Susan Marie. "Fantasy Proneness, Amnesia, and the UFO Abduction Phenomenon." *Dissociation* IV, no. 1 (1991): 46-54.

Proud, Louis. "Aliens, Predictions & the Secret School: Decoding the Work of Whitley Strieber." New Dawn, accessed August 18, 2018, https://www.newdawnmagazine.com/articles/aliens-predictions-the-secret-school-decoding-the-work-of-whitley-strieber.

Radford, Benjamin. "Astral Projection: Just a Mind Trip." Live Science, 2017, accessed June 9, 2018, https://www.livescience.com/27978-astral-projection.html.

Radin, Dean. "Thinking About Telepathy." *Think* 1, no. 3 (2003): 23-32.

Raynes, Brent. "Trying to Understand the Psychic Mind." AP Magazine, accessed December 27, 2018, http://www.apmagazine.info/index.php?option=com_content&view=article&id=388:reality-checking-june-2013&Itemid=194.

"Reality Checking." AP Magazine, 2011, accessed December 26, 2018, http://www.apmagazine.info/index.php?option=com_content&view=article&id=1152:reality-checking-june-2018&catid=2&Itemid=44.

Ring, Kenneth. "The Omega Project: Near-Death Experiences, UFO Encounters, and Mind at Large." The Intuitive-Connections Network, accessed June 24, 2018, http://www.intuitive-connections.net/2008/book-omegaproject.htm.

Ring, Kenneth, and Sharon Cooper. *Mindsight: Near-Death and Out-of-Body Experiences in the Blind.* Bloomington, IN: iUniverse Inc., 2008.

Ring, Kenneth, and Christopher Rosing. "The Omega Project: An Empirical Study of the NDE-Prone Personality." *Journal of Near-Death Studies* 8, no. 4 (1990): 211-38.

Ring, Kenneth, and Evelyn Elsaesser Valarino. *Lessons from the Light: What We Can Learn from the near-Death Experience.* Needham MA: Moment Point, 2006.

Rivera, Heather. "Healing the Present from the Past." PhD Dissertation.

Rivera, Heather S. Friedman. *Healing the Present from the Past: The Personal Journey of a Past-Life*

*Researcher.* Bloomington, IN: Balboa Press, 2012.

Robey, Michael J. "Send Love Telepathically." Psychic Tarot Oracle Reading and Courses, accessed December 17, 2017, https://www.psychicgr.com/post/2017/01/16/send-love-telepathically.

Robey, Michael J. "So You Want to Be a Channeler Too?" SelfGrowth.com, accessed March 1, 2019, https://www.selfgrowth.com/articles/so-you-want-to-be-a-channeler-too.

Rodwell, Mary. *Awakening: How Extraterrestrial Contact Can Transform Your Life.* New Mind Publishers, 2010.

"Science and the Near-Death Experience." Near-Death Experiences and the Afterlife, accessed March 18, 2019, https://www.near-death.com/science/research/science.html.

"Science and the near-Death Experience." Near-Death Experiences and the Afterlife, accessed March 4, 2019, https://www.near-death.com/science/research/science.html.

"Science Quotes by Albert Einstein." Todayinsci, accessed March 9, 2018, https://todayinsci.com/E/Einstein_Albert/EinsteinAlbert-Quotations.htm.

Seaburn, Paul. "UFO Studies to Be Taught at Major University." Mysterious Universe, 2018, accessed January 15, 2019, https://mysteriousuniverse.org/2018/12/ufo-studies-to-be-taught-at-duke-university/.

"Secrets of Learning." Altered States, accessed May 15, 2018, https://docs.google.com/document/d/1w5SYjQcCQBDea8y6pG7L8fz33cNoP8FG-PYEDGx9oyI/edit?ts=5ae7a36f.

Singh, Jupinderjit. "Can Science Uphold the Belief in Rebirth." *The Indian Tribune,* 2002, accessed April 5, 2018, https://www.tribuneindia.com/2002/20020622/.

"Skeptical Argument: Nothing Useful Comes from

NDEs." Near-Death Experiences and the Afterlife, accessed May 23, 2018, https://www.near-death.com/index.html.

"Sleep Paralysis, Vibrations, Spiritual Revelation, and Alien Abduction or Getting out of 4d," *Fahrusha's Weblog*, October 16,

2010, https://fahrusha.wordpress.com/2010/02/27/sleep-paralysis-vibrations-spiritual-revelation-and-alien-abduction-or-getting-out-of-4d/.

"Socrates (469–399 B.C.E.)." Internet Encyclopedia of Philosophy, accessed September 30,

2018, https://www.iep.utm.edu/socrates/.

"Socrates Quotes." Goodreads, accessed March 2,

2019, https://www.goodreads.com/quotes/113638-the-greatest-blessing-granted-to-mankind-come-by-way-of.

Spanos, Nicholas, Patricia Cross, Kirby Dickson, and Susan Dubreuil. "Close Encounters: An Examination of UFO Experiences." *Journal of Abnormal Psychology* 102, no. 4 (1993): 624-32.

"Spontaneous Healings." Eternea, accessed September 18,

2018, https://www.eternea.org/spontaneous_healings.aspx.

Starr, Michelle. "Here's How We Might Find a Wormhole, if They Existed." *Science Alert.* (2019) accessed March 01. https://www.sciencealert.com/wormholes-probably-don-t-exist-but-here-s-how-we-could-find-one-if-they-did.

Steiger, Brad, and Sherry Hansen Steiger. *The Gale Encyclopedia of the Unusual and Unexplained.* USA: Gale Group Inc., 2003.

Sunfellow, David. "Miraculous NDE Healings." The Formula for

Creating Heaven on Earth, 2019, accessed February 15, 2019, https://the-formula.org/miraculous-nde-healings/.

"Survey Results." Astralinfo.org, accessed July 18, 2018, http://www.astralinfo.org/survey-results/.

"Susan Blackmore." Wikiquote, accessed February 20, 2019, https://en.wikiquote.org/wiki/Susan_Blackmore.

Talbott, David, and Wallace Thornhill. "Excerpts from Thunderbolts of the Gods." The Thunderbolts Project, accessed July 29, 2017, http://www.thunderbolts.info/wp/resources/thunderbolts-of-the-gods-excerpts/.

Tart, Charles T. "A Future for Dualism as an Empirical Science?" New Dualism Archive, 2006, accessed December 17, 2018, http://www.newdualism.org/papers/C.Tart/Tart-Dualism.pdf.

"Telepathy and the near-Death Experience." Near-Death Experiences and the Afterlife, accessed March 8, 2018, https://www.near-death.com/science/research/telepathy.html.

"The Day My Life Changed: February 2, 2006." Anita Moorjani Love is Not A Luxury, accessed February 2, 2019, https://anitamoorjani.com/about-anita/near-death-experience-description/.

"The Filter Factor." Astral Voyage, accessed November 20, 2018, https://www.astralvoyage.com/astral-projection/filter-factor.php.

"The Physiology of Psychic People and Psychic Ability." Core Spirit, accessed May 4, 2018, https://weilerpsiblog.wordpress.com/2012/08/11/the-physiology-of-psychic-people-and-psychic-ability/.

"The Infrasound of a Forest Elephant Rumble." The Cornell Lab of Ornithology, accessed September 4, 2017, http://www.birds.cornell.edu/brp/elephant/cyclotis/language/infrasound.html.

"The Science of Life Discovered from Lynnclaire Dennis' NDE." Near-Death Experiences and the Afterlife, accessed January 23, 2018, https://www.near-death.com/science/articles/science-of-life-from-an-nde.html.

"The Vestibular System: Definition, Anatomy, and Function." accessed May 12, 2018, https://study.com/academy/lesson/the-vestibular-system-definition-anatomy-function.html.

"This Is Exactly Why Nikola Tesla Told Us to Study the 'Non Physical.'" Cosmic Scientist, 2016, accessed January 20, 2019, http://www.cosmicscientist.com/this-is-exactly-why-nikola-tesla-told-us-to-study-the-non-physical/.

"Thomas Sawyer's Near-Death Experience." Near-Death Experiences and the Afterlife, accessed November 25, 2018, https://www.near-death.com/reincarnation/experiences/thomas-sawyer.html.

"Thomas Szasz." Wikiquote, accessed March 15, 2019, https://en.wikiquote.org/wiki/Thomas_Szasz.

"Tom Wilson Quotes." BrainyQuote, accessed March 21, 2019, https://www.brainyquote.com/authors/tom_wilson.

"Top Universities in the World 2019." Topuniversities, 2019, accessed February 10, 2019, https://www.topuniversities.com/university-rankings-articles/world-university-rankings/top-universities-world-2019.

"Transcendental Omnipresence." Of Quasars and Quanta, accessed December 12, 2018, http://www.cosmiclight.com/ofquasars/omnipresence.html.

Tucker, Jim. "Advice to Parents of Children Who Are Spontaneously Recalling Past Life Memories." Division of Perceptual Studies, accessed May 17, 2018, https://med.virginia.edu/perceptual-studies/resources/advice-to-parents-of-children-who-are-spontaneously-recalling-past-life-memories/.

Twemlow, Stuart. "Clinical Approaches to the Out-of-Body Experience." *Journal of Near-Death Studies* 8, no. 1 (1989): 29-43.

Twemlow, Stuart W., Glenn O. Gabbard, and Fowler C.

Jones. "The Out-of-Body Experience: A Phenomenological Typology Based on Questionnaire Responses." *American Journal of Psychiatry* 139, no. 4 (1982): 450-55.

"UFO Sightings by Astronauts." NOUFORS, 1993, accessed May 9, 2018, http://noufors.com/ufo_sightings_by_astronauts.html.

Van Lommel, Pim. *Consciousness Beyond Life: The Science of the Near-Death Experience.* New York, NY: HarperCollins, 2010.

"Vibrational State." Astralinfo.org, accessed August 7, 2018, http://www.astralinfo.org/?s=vibration.

"W.B. Yeats Quotes." Goodreads, accessed January 9, 2019, https://www.goodreads.com/quotes/122468-the-world-is-full-of-magic-things-patiently-waiting-for.

Wade, Jenny. "The Phenomenology of Near-Death Consciousness in Past-Life Regression Therapy: A Pilot Study." *Journal of Near-Death Studies* 17, no. 1 (1998): 31-53.

Waggoner, Robert. *The Lucid Dreaming Pack: Gateway to the Inner Self.* San Francisco, CA: Moment Point Press, 2009.

Wagner, Stephen. "Is There Evidence of Reincarnation?" ThoughtCo., 2018, accessed September 4, 2018, https://www.thoughtco.com/reincarnation-best-evidence-2593151.

"What Are Dissociative Disorders?" American Psychiatric Association, 2016, accessed May 28, 2018, https://www.psychiatry.org/patients-families/dissociative-disorders/what-are-dissociative-disorders.

"What Are Orbs?" Orb Whisperer, accessed January 20, 2019, https://orbwhisperer.com/what-are-orbs%3F.

"What Does It Mean to Be a Highly Sensitive Person?" PsychCentral, 2018, accessed June 20, 2018, https://psychcentral.com/lib/what-does-it-mean-to-be-a-highly-sensitive-person/.

Weiss, Brain. *Many Lives, Many Master.* New York, NY: Simon & Schuster, 1988.

Weiss, Brian. "Many Lives, Many Masters." 1988, accessed February 6, 2019, http://siqv.weebly.com/uploads/2/3/5/1/23516512/many_lives_many_masters_-_brian_weiss.pdf.

"Welcome to Dolores Cannon.Com." Dolores Cannon, accessed March 23, 2018, https://dolorescannon.com/.

Wenner, Melinda. "Belief in Reincarnation Tied to Memory Errors." NBC News, 2007, accessed May 26, 2018, http://www.nbcnews.com/id/17982545/ns/technology_and_science-science/t/belief-reincarnation-tied-memory-errors/#.XFZXfM9Khje.

"What Happens to the Fabric of Space-Time When an Object Moves through it near the Speed of Light?" Gravity Probe B: Testing Einstein's Universe, accessed March 3, 2019, https://einstein.stanford.edu/content/relativity/q909.html.

"What Is Matter? —Never Mind. What Is Mind? —No Matter." "Punch," London humor magazine, accessed February 23, 2019, https://quoteinvestigator.com/2018/08/30/matter-mind/.

"Who Is Saying It? – Quotes." Youareme.com, accessed February 16, 2019, http://www.youareme.com/quotes.html.

"Why Near-Death Experiences Are Not Hallucinations." Near-Death Experiences and the Afterlife, accessed January 28 2019, https://www.near-death.com/science/hallucinations/why-ndes-are-not-hallucinations.html.

"William Lawrence Bragg." Goodreads, accessed March 12, 2019, https://www.goodreads.com/quotes/166826-the-important-thing-in-science-is-not-so-much-to.

"William Ralph Inge Quote." BrainyQuote, accessed

December 10, 2018, https://www.brainyquote.com/quotes/william_inge_1 49271.

Williams, Sarah C. P. "Sounds You Can't Hear Can Still Hurt Your Ears." *Science*, 2014, accessed January 11, 2018, https://www.sciencemag.org/news/2014/09/sounds-you-cant-hear-can-still-hurt-your-ears.

Woolger, Roger J. "Past Life Therapy, Trauma Release and the Body." Carol Bowman, Past Life Therapy, accessed February 6, 2019, https://www.carolbowman.com/library/past-life-therapy-trauma-release-body.

"World University Rankings." Top Universities, 2019, accessed February 10, 2019, https://www.topuniversities.com/university-rankings/world-university-rankings/2019.

# About Atmosphere Press

Atmosphere Press is an independent, full-service publisher for excellent books in all genres and for all audiences. Learn more about what we do at atmospherepress.com.

We encourage you to check out some of Atmosphere's latest releases, which are available at Amazon.com and via order from your local bookstore:

*Geometry of Fire,* nonfiction by Paul Warmbier

*Chasing the Dragon's Tail,* nonfiction by Craig Fullerton

*Pandemic Aftermath: How Coronavirus Changed Global Society,* nonfiction by Trond Undheim

*Great Spirit of Yosemite: The Story of Chief Tenaya,* nonfiction by Paul Edmondson

*My Cemetery Friends: A Garden of Encounters at Mount Saint Mary in Queens, New York,* nonfiction and poetry by Vincent J. Tomeo

*Change in 4D,* nonfiction by Wendy Wickham

*Disruption Games: How to Thrive on Serial Failure,* nonfiction by Trond Undheim

*Eyeless Mind,* nonfiction by Stephanie Duesing

# About Lynn Miller

Lynn holds dual BS degrees in Psychology and Biology, and a MS in Biology. For several years, she worked in the food industry as a microbiologist. Lynn served as an Adjunct Professor at Pensacola State College, where she taught Botany,  Microbiology, and Biology. She has taught High School Biology and art, k-12 for thirteen years. Lynn is a frequent co-host with Brent Raynes, Alternate Perception Audio Interview Series. Influenced by the work of William Buhlman, Lynn has practiced controlled out-of-body experiences since 2009. For over fifteen years, she has extensively researched consciousness.

# About Barbara Mango

Barbara received her MA and Ph.D. in Metaphysics. Prior to receiving her post graduate degrees, she worked as a pre-K educator and Reiki practitioner. She is the sole proprietor of *A Healing Touch Reiki*. Barbara currently researches, writes, and speaks about extraordinary phenomena  and consciousness. Barbara is a Board Member of the PLR Institute (Past Life Research Institute). She also served on the research committee for the Dr. Edgar Mitchell Foundation. Barbara co-hosts Alternate Perception Radio with Brent Raynes (www.apmagazine.info). *Convergence* is her debut book. Additionally, she is a contributing author to the book, *The Transformative Power of Near-Death Experiences*, by Dr. Penny Sartori (Watkins, 2017). When not writing, Barbara can be found with her nose in a book, puttering in her garden, or playing with her three neurotic cats. She lives in the Northeast with her husband. She may be contacted at her website, www.extraordinaryexperiences.org
or on facebook.com/journeysintoconciousness

CPSIA information can be obtained
at www.ICGtesting.com
Printed in the USA
BVHW071228220221
600770BV00007B/508